Mathematical Mysteries in the Natural World

Why the Small Outnumbers the Big

Mathematical Mysteries in the Natural World

Why the Small Outnumbers the Big

Alex Ely Kossovsky

World Scientific

NEW JERSEY · LONDON · SINGAPORE · BEIJING · SHANGHAI · HONG KONG · TAIPEI · CHENNAI · TOKYO

Published by

World Scientific Publishing Co. Pte. Ltd.

5 Toh Tuck Link, Singapore 596224

USA office: 27 Warren Street, Suite 401-402, Hackensack, NJ 07601

UK office: 57 Shelton Street, Covent Garden, London WC2H 9HE

Library of Congress Cataloging-in-Publication Data

Names: Kossovsky, Alex Ely, author.
Title: Mathematical mysteries in the natural world : why the small outnumbers the big /
    Alex Ely Kossovsky.
Description: New Jersey : World Scientific, [2026] | Includes bibliographical references and index.
Identifiers: LCCN 2025007457 | ISBN 9789819801824 (hardcover) |
    ISBN 9789819803019 (paperback) | ISBN 9789819801831 (ebook) |
    ISBN 9789819801848 (ebook other)
Subjects: LCSH: Benford's law (Mathematics) | Numerical analysis.
Classification: LCC QA273.6 .K673 2025 | DDC 518--dc23/eng/20250625
LC record available at https://lccn.loc.gov/2025007457

British Library Cataloguing-in-Publication Data

A catalogue record for this book is available from the British Library.

For any available supplementary material, please visit
https://www.worldscientific.com/worldscibooks/10.1142/14070#t=suppl

Desk Editors: Aanand Jayaraman/Gabriel Rawlinson

Typeset by Stallion Press
Email: enquiries@stallionpress.com

About the Author

Alex Ely Kossovsky is the author of the books *Benford's Law: Theory, the General Law of Relative Quantities, and Forensic Fraud Detection Applications*, World Scientific Publishing Company, 2014; *Studies in Benford's Law: Arithmetical Tugs of War, Quantitative Partition Models, Prime Numbers, Exponential Growth Series, and Data Forensics*, Kindle Direct Publishing, Amazon, 2019; *The Birth of Science*, regarding Kepler's Celestial Data Analysis, Galileo's Terrestrial Experiments, and Newton's Grand Synthesis, Springer Nature Publishing, 2020; and *A Comprehensive Summary of the Benford's Law Phenomenon: On the Unequal Spread of Digits within Scientific and Typical Data*, World Scientific Publishing Company, 2025. Kossovsky is the inventor of a patented mathematical algorithm in data fraud detection analysis, registered at the US Patent Office. The author specialized in Applied Mathematics and Statistics at the City University of New York, and in Physics and Pure Mathematics at the State University of New York at Stony Brook. Email address: akossovsky@gmail.com.

Contents

Section 2: Causes and Explanations of the Phenomenon   **109**

Section 4: Benford's Law as a Direct Consequence of the General Law

Introduction

The excessive and exaggerated New Year celebrations on the last day of December and at the beginning of 2004 soon gave way to a more subdued and darker public mood as two brutal wars were raging simultaneously in West and Central Asia, while the harsh New York City winter was just beginning in earnest, wearing people down with heavy snow and unusually cold sub-zero temperatures. This author, while far away from New York during that January, was casually scanning from afar the set of mathematics and statistics courses offered on the website of the City University of New York for the upcoming spring semester, when he was suddenly jolted into a thrilling and energetic mood, noticing one unique course being offered there titled "Data Analysis". Lost in thought and oblivious to all his surroundings, he whispered to himself: "What on Earth could be taught in such a course?" He was wondering if it might include some exotic, new, and more potent regression and correlation methods, perhaps linking entities and variables that, on the face of it, appear totally unrelated and unconnected. Also, the possibility that some data fraud detection methods and novel auditing algorithms might be included in the course to expose fake data or manipulated numbers, detecting financial corruption, election vote theft, or fraudulent scientific publications based on false and dishonest experimental data — all enticed the author even more to register for it.

The course was taught by the chairman of the Applied Mathematics and Statistics Department himself, the late Distinguished Professor Edward Binkowski, a protégé (akin to a biblical disciple, rather) of John Tukey of Princeton University — widely considered as one of the most significant statisticians of the post-war era, known for coining the term "software" and the concept of "bit", with contributions spanning various areas, including data analysis, time series analysis, and the development of statistical computing. The course indeed turned out to be an extraordinary educational experience for the author. Binkowski taught with the utmost dedication to the mathematical and scientific truth, with an open-minded and flexible approach, often thinking outside the box, inserting skepticism and demanding further inquiry and scrutiny wherever he deemed it necessary.

Ed Binkowski as a Princeton University Alumnus in 1974

It was then, during one of Binkowski's captivating evening classes in March of 2004, that I very briefly heard about the Benford digital phenomenon for the first time ever, although this was not presented by him as a quantitative phenomenon in favor of the small at the expense of the big, but rather as a supposedly purely digital and numerical phenomenon. Binkowski interrupted his normal schedule and the usual planned presentation of the class curriculum and announced: "You students should know that most numbers in the real world start with low digits, 1, 2, or 3, while only a minority of numbers start with the higher digits 4 to 9". Listening to him stating this caused me to feel a great deal of cognitive discomfort, surprise, and a certain rebelliousness. Was I hearing him correctly? And if so, did he really mean what he said, or was he just mixing or confusing his words in error, uttering his sentence without enough focus? Isn't this against all intuition and common sense? Shouldn't numbers that arise from chaotic and random real-world physical entities yield a totally random and fair distribution of digits (an equal spread of all possible digits)? Instantly I raised my hand, and politely but in a slightly impatient tone of voice I asked: "Professor, could you repeat what you said please?" Binkowski smiled broadly, appearing quite delighted, perhaps because my questioning him made him realize that at least one student was really paying attention to what he was saying in class, or maybe because after already knowing me for a short while he was expecting the statement to puzzle me and that I would thus reject it out of hand. And so, with a joyful expression on his face, he retorted "Yes mister Kossovsky, let me repeat, digits in the first position in numbers do not occur evenly in the real world of data, rather low digits strongly dominate high digits there, and this is called Benford's Law, so while I understand that this is quite surprising and counterintuitive for you, yet this is a fact of life, empirically confirmed in nearly all data types; therefore, you better start internalizing and accepting all this!" I was too busy during that semester and even during the rest of the year to look into this strange law of numbers or to try to think of an explanation, but it stayed in the back of my mind as something annoying that demanded attention, something that I would surely need to investigate later.

It was only much later, during the next winter break of January 2005, with plenty of time on hand, that I seriously sat down and attempted to simulate and calculate imaginary data sets in the abstract and investigate their resultant phantom digital configurations, all of which led me to the discovery of the chains of statistical distributions and their strong tendency to produce many small quantities but only very few big ones, while simultaneously leading also to Benford digital behavior. Binkowski urged me to publish my results immediately, and then he gradually became my mentor, supporting, encouraging, and helping in every which way necessary. Sadly, Binkowski passed away in 2019, leaving behind numerous students and colleagues who will always remember him as a kindhearted and honest person, full of humanity, and with a great sense of humor. Binkowski was an exceptionally well-read academic and a true intellectual with broad interests, thought of by many as a "renaissance man" of sorts. His large apartment in the bohemian and artistic Greenwich Village neighborhood of New York City, loaded with thousands of books on numerous shelves covering nearly all the walls, appeared more like a library than a personal apartment.

The physical world around us is quantitatively structured in a very particular way, so that the vast majority of entities are small, while only a tiny minority are big. There exist very few exceptions to this nearly universal rule. It is not only that we humans tend to build small factories, houses, cars, tables, dishes, and cakes, as opposed to super big things, but this is also the way things are naturally produced and created in the physical world and throughout the entire universe. Mother Nature is always quite busy, simultaneously forming planets, stars, galaxies, rivers, cities, and towns. Yet, in spite of her hurried and difficult schedule she is very picky when it comes to quantitative style, and so she takes her time, deliberately creating things her way. Even in her rare moments of anger, when she rattles planets with earthquakes and erupts volcanoes, even when she rages and spectacularly lightens the sky with supernova explosions, causing the death of huge stars, smashing them to smithereens, she still exercises some self-control and pauses a moment to calculate the resultant relative quantities carefully, so as to create or destroy always in her own particular way. She greatly favors and sympathizes with the small and the weak, and therefore she creates many of them, while disliking and suspecting the big and the powerful, producing very few of them. The prediction of Benford's Law regarding the frequent occurrences of low digits such as {1, 2, 3} in the first position in numbers, on the left-most side, versus the rarity of high digits such as {7, 8, 9}, is simply a consequence and the result of the original and primordial quantitative rule in the physical world favoring the small over the big. Professor Binkowski did not know that for this entire phenomenon the quantitative drives the digital and the numerical, until the author clearly demonstrated it, and finally Binkowski adopted the author's point of view and predicted that the quantitative law termed as GLORQ in this book will eventually be unreservedly recognized and approved by future generations of mathematicians.

SECTION 1
Empirical Evidence of the Small Is Beautiful Phenomenon

Chapter 1

The Small Is Beautiful Phenomenon

Most people believe or assume that small things are more numerous in the world than big things. Surely there are more villages than towns, more towns than cities, and more cities than metropolises. Surely there are more raindrops than puddles, more puddles than lakes, and more lakes than seas or oceans. The principle appeals to common sense. Typically for older and experienced statisticians, scientists, and professionals, this conviction arises consciously or subconsciously from numerous real life experiences and actual numerical analyses with data on physical phenomena. For others, this belief arises more from general intuition or some very vague reasoning, and thus when pressed for an explanation, they cannot explicitly express any concrete argument in support of this notion, and none can come up with any generic mathematical or statistical proof in favor of the small.

It appears that nobody in the literature has ever paid much attention to this neglected yet important topic regarding the occurrences of relative quantities in the world; and certainly nobody has ever attempted to find an exact numerical pattern for the phenomenon. The goal of this book is not merely to present a rich variety of the manifestation of the phenomenon and provide three generic explanations for it, but also to arrive successfully at an exact numerical pattern indicating by how much the small is more numerous than the big. This numerical size pattern turned out to be nearly universal, as it is found in almost all data sets relating to physics, chemistry, astronomy, geology, biology, engineering, economics, finance, accounting, governmental census

3

and demographic information, various statistical distributions, combinatorics, integer partitions, and numerous mathematical sequences, sets, and algorithms.

To name just one example in the context of pure mathematics and abstract numbers, the finite Fibonacci Sequence limited to the maximum value of N contains by far many more small values than medium values, and it also contains by far many more medium values than big values. The terms 'small', 'medium', and 'big' refer to **relative quantities** within the framework of the entire data set under consideration, which for the finite Fibonacci Sequence ranges from 1 to N. These terms regarding sizes never refer to any **absolute quantities** or some imaginary fixed and universal benchmark values applicable to all existing data sets. For example, it would be ludicrous to designate 0.001 as the absolute small and 10^{13} as the absolute big, pertaining to all data sets. Sizes are determined locally relative to the data set itself, not globally relative to some imaginary universal benchmarks.

Let us present a simple argument with regards to data sets pertaining to physical man-made entities, purporting that in general people tend to produce many more small things than big things. This is done by examining several common human activities, all of which strongly confirm the principle.

It is much more common and a lot easier to collect and smelt small pieces of gold for earrings, rings, or bracelets, than for big 12.4 kilogram (438.9 ounce) standard gold bars; and indeed there is by far much more small gold jewelry in the world than big gold bars such as those held and traded internationally by central banks and bullion dealers. It is much easier to produce a small personal computer than a big super computer; and indeed there are by far many more small personal computers in the world than big super computers. It is much easier to build a small building or a small house than a big skyscraper; and indeed there are by far many more small houses and buildings in the world than big skyscrapers. Figures 1.1–1.3 dramatize the distinct sizes for these three human activities.

It might be argued that these three examples above are all related to human effort and actions, such as smelting gold, producing computers, and building houses and skyscrapers, hence perhaps the small is more numerous simply because we prefer small things for

Figure 1.1: Many More Small Earrings, Rings, and Bracelets Are Made than Big Gold Bars

Figure 1.2: Many More Small Personal Computers Are Produced than Big Super Computers

one reason or another, or that we are not often in need of big things. Perhaps big things such as gold bars, super computers, and tall buildings are too expensive to obtain and to build. Perhaps we tend to become tired and weary after hard labor, thus we typically produce small things. In other words, we tend to avoid creating big things which are more difficult to do and are more time consuming,

Figure 1.3: Many More Small Houses and Buildings Are Constructed than Big Skyscrapers

unless we are truly compelled by circumstantial necessity, or unless we become overly ambitious and enthusiastic in rare moments of great inspiration.

The argument above mistakenly limits the phenomenon to human related constructs, yet in fact, the manifestation of the phenomenon has a much wider scope than merely human activities. When natural phenomena are considered, and when entities that exist totally independently of us and our activities are quantitatively analyzed, it is found that the small almost universally outnumbers the big. Even though Mother Nature does not seem to ever get tired or weary, and she certainly would never worry about spending a lot of time and effort in building ever bigger things, yet she has the same tendencies as humans do, and she strongly favors the small over the big.

What is behind such nearly universal pattern? Are there generic causes in nature that produce such uneven and skewed occurrences in relative quantities? Is this a secondary physical law of nature which can be derived from the primary laws of physics, or is it simply a statistical phenomenon? In addition, one wonders why this exact size pattern appears also in pure mathematical constructs — totally unrelated to natural phenomena or human activities.

The expression or motto in this book for this phenomenon regarding size configuration is coined **small is beautiful**. The term 'beautiful' in this context is not meant literally, but rather metaphorically, as it signifies the connotation associated with the adjectives favored, favorite, numerous, plentiful, frequent, most common, and abundant.

Chapter 2

Small Life Forms Decisively Outnumber Big Creatures

An instructive demonstration of the small is beautiful phenomenon in the physical sciences is found in biological data, relating body size with population size (abundance). The bigger the creature the fewer there are of it in the world. The smaller the creature the more there are of it in the world. As an example, there are relatively few big whales swimming the oceans; only approximately 1.5 million such individuals, but there are relatively quite numerous small birds flying the skies; approximately 300 billion individuals. Hence, there are approximately 200,000 small birds in the skies for each single big whale in the oceans!

We admire the whales for their intelligence, complexity, and size, as well as for their amazing successful return to the sea after becoming mammals on land. We admire the birds for being perhaps the first to institute the ('happy') marriage, and for the cooperative and strong efforts of the two caring and loyal parents in providing food and protection for their young and vulnerable offsprings.

The table depicted in Figure 1.4 clearly demonstrates the obvious relationship between creature size and its global population count. Values in the table are not exact of course, but rather very good estimates by way of searching reliable sources of information in biology. Difficulties arise from the fact that there exist different species of the same creature with differing body length, nonetheless information about global population size is usually provided in extreme generality encompassing all species. For example, there exist

CREATURE	METERS	POPULATION
Bacteria	0.000001	5,000,000,000,000,000,000,000,000,000,000
Mosquitoes	0.004	70,000,000,000,000,000
Ants	0.007	1,000,000,000,000,000
Birds	0.20	300,000,000,000
Humans	1.65	7,000,000,000
Whales	20	1,500,000

Figure 1.4:   Small Life Forms Decisively Outnumber Big Creatures

many species of ants with varying sizes, such as the South American giant Amazonian ants with about 3–4 cm length, as well as the workers of the little black ant *Monomorium minimum* in North America with only about 1–2 mm length. Yet this challenge is easily overcome by averaging out the sizes for a variety of different species, and the final conclusion about the abundance of the small and the scarcity of the big is not affected at all by such approximations.

Figure 1.5 depicts the bar chart summarizing the data of the table in Figure 1.4. This is not exactly a bar chart since it shows comparisons among discrete categories which are aligned and ranked numerically; rather this is close to being a histogram of sorts. Histograms count the number of cases or occurrences falling within sub-intervals of the entire range, but here no such counting is taking place. We never count here for example the population of all creatures between 1 centimeter and 4 centimeters. Instead we focus on a specific creature and count its population. Nonetheless, it is almost a certainty that a comprehensive pan-biological histogram encompassing all creatures would show the same pattern of skewness, falling to the right, and mimicking the shape seen in Figure 1.5. The construction of such a huge and detailed histogram would surely demand the collaboration of many biologists, and would be an extremely challenging and time consuming endeavor.

In addition, it should be noted that the horizontal x-axis is not drawn according to any consistent scale at all, rather it is a hybrid and distorted scale of sorts, mixing the meter (symbol m) with the millimeter (0.001 of a meter — symbol mm), as well as the micrometer (0.000001 of a meter — symbol μm). Nonetheless, the values marked on the horizontal scale increase monotonically

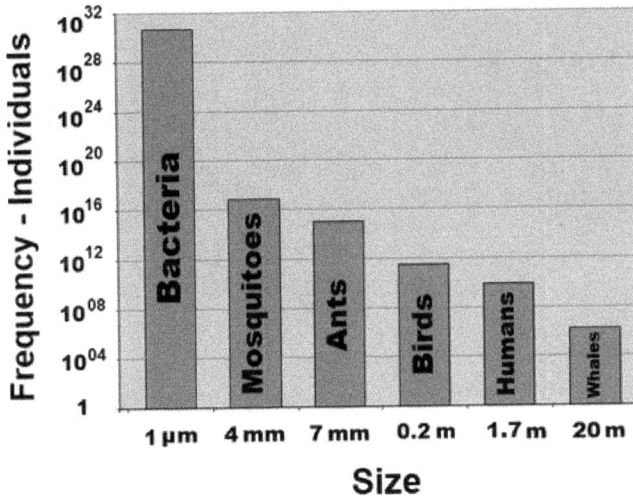

Figure 1.5: Bar Chart of Population by Size for Six Life Forms and Creatures

and consistently to the right in the same manner as do all proper histogram scales.

The vertical axis on the other hand is a proper and consistent scale, although it uses the logarithmic scale, so that going from 1 to 10000 appears just as long as in going from 10000 to 100000000, and just as long as in going from 100000000 to 1000000000000; and in spite of the fact that distances between 1, 10000, 100000000, and 1000000000000 are dramatically different. Logarithmic scale compresses different segments of the scale in certain ways so that the data in its entirely can be clearly visualized. If not for the use of the logarithmic scale here, the big would not even be seen in the chart at all since it is so insignificant compared with the small.

The logarithmic scale uses constant and equal spacing between the values of 1, 10, 100, 1000, and so forth; namely between 10^0, 10^1, 10^2, 10^3, and so on, and it is typically drawn with a line for each integral power of 10. Figure 1.5 though shows only a single line for each 4th exponent.

Chapter 3

Small Molecules Outnumber Big Molecules in the Chemical World

A large list of 2175 commonly used and naturally occurring chemical compounds is provided on the website: http://www.convertunits. com/compounds/. The website criterion for the selection of these 2175 molecules does not follow any particular formal procedure, and instead it simply pulls together information from a variety of chemical and scientific sources, as well as utilizing the informal suggestions of chemists and scientists regarding which molecules should be considered as relevant and important for compilation. The large variety in this list confers to it legitimacy as a good representative of the use and occurrences of chemical molecules in the world for the purpose of examining relative quantities. The list encompasses chemicals relating to the pharmaceutical industry, metallurgical plants, food industry, heavy industry, and others. It lists synthetic (laboratory-made) molecules as well as naturally occurring compounds. Since the selection criterion does not involve the molar mass (molecular mass) in any way, the weight of the molecules can then be thought of as a truly random variable in this list.

The following seven compounds are selected as examples from the long list of 2175 molecules, including the smallest molecule lithium hydride, and the biggest molecule bismuth subnitrate:

Lithium hydride	LiH	molar mass = 7.9489 gram/mol
Water	H_2O	molar mass = 18.0153 gram/mol
Hydrogen peroxide	H_2O_2	molar mass = 34.0147 gram/mol
Sodium nitrate	$NaNO_3$	molar mass = 84.9947 gram/mol
Glucose	$C_6H_{12}O_6$	molar mass = 180.1559 gram/mol
Mercury sulfate	$HgSO_4$	molar mass = 296.6526 gram/mol
Bismuth subnitrate	$Bi_5O(OH)_9(NO_3)_4$	molar mass = 1461.9870 gram/mol

Examination of the data reveals that the distribution of molar mass is highly skewed in favor of the small. Figure 1.6 depicts the histogram up to 900 gram/mole. Clearly small molecules outnumber big molecules in general, except for a brief rise on the very left of the histogram between 0 and 300 where relatively bigger molecules are slightly more numerous than smaller molecules. Such temporary and minor reversal of the phenomenon in the beginning for very low values is quite typical in many other physical, scientific, financial, and accounting data sets, and this slight antithesis feature within the overall small is beautiful phenomenon shall be encountered again in the chapters regarding astronomical data on pulsar rotation rate and revenue accounting data.

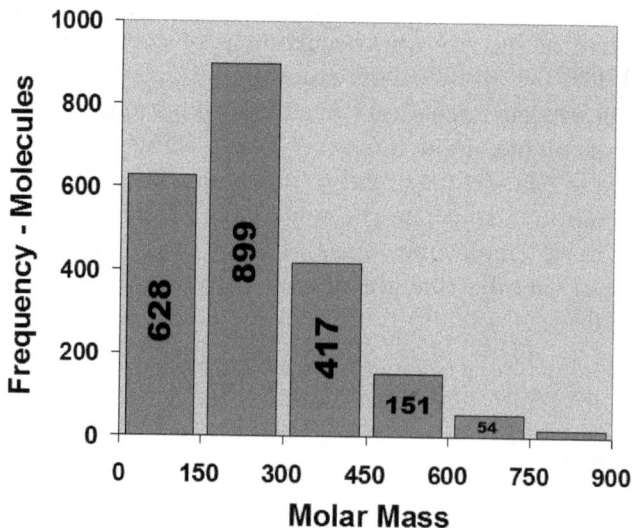

Figure 1.6: Small Molecules Generally Outnumber Big Molecules

The tail to the right of Figure 1.6 is actually a bit longer. The complete distribution is:

There are	**628**	molecules with molar mass between	0	and	150.
There are	**899**	molecules with molar mass between	150	and	300.
There are	**417**	molecules with molar mass between	300	and	450.
There are	**151**	molecules with molar mass between	450	and	600.
There are	**54**	molecules with molar mass between	600	and	750.
There are	**16**	molecules with molar mass between	750	and	900.
There are	**3**	molecules with molar mass between	900	and	1050.
There are	**5**	molecules with molar mass between	1050	and	1200.
There are	**0**	molecules with molar mass between	1200	and	1350.
There are	**2**	molecules with molar mass between	1350	and	1500.

Chapter 4

Small Atoms Outnumber Big Atoms in the Composition of Chocolate

When chocolate is chemically analyzed in terms of its atomic components, it is found that small atoms are highly abundant, while big atoms are extremely rare. It should be emphasized that this component analysis is not done along compounds/molecules, but rather along basic atoms.

Estimation is done by first breaking down chocolate into its typical ingredients, such as milk (9.5%), sugar (50%), cocoa mass (30%), cocoa butter (10%), and lecithin (0.5%), and assuming the above relative concentration by weight of each ingredient. *Milk* is analyzed in terms of its constituents, such as protein, saturated fatty acids, monounsaturated fatty acids, lactose, cholesterol, calcium, and so forth. *Sugar* is approximated as the molecule glucose $C_6H_{12}O_6$. *Cocoa mass* is analyzed in terms of its constituents, such as carbohydrates, fat, protein, calcium, iron, magnesium, manganese, phosphorus, potassium, zinc, and so forth. *Cocoa butter* is approximated as the molecules palmitic acid $C_{16}H_{32}O_2$ (28%), oleic acid $C_{16}H_{34}O_2$ (36%), and stearic acid $C_{18}H_{36}O_2$ (36%). *Lecithin* is simply $C_{35}H_{66}NO_7P$.

Secondly, a count is made of all the molecules according to their frequency of occurrences within the constituents of all five ingredients. Thirdly, each such molecule is analyzed in terms of its constituent atoms, and a final count is made of each atom. It should be noted that even though this is a very rough estimate, and the exact count of atoms here is simply the result of very crude approximations, the conclusion regarding the manifestation of the small is beautiful phenomenon is so decisive and dramatic here that even fairly large errors in the approximations would not sway the final overall conclusion.

The estimated result for a minuscule piece of chocolate weighing 0.000000000000684 grams shows that it is made almost exclusively of hydrogen, carbon, nitrogen, and oxygen atoms, as shown in Figure 1.7. In fact, in terms of a comparative count of atoms, these four small atoms together constitute 99.8% of chocolate, while the bigger atoms constitute only 0.2% of it.

Hence, this analysis clearly demonstrates that chocolate is almost entirely made of small atoms having atomic numbers between 1 and 8; and that big atoms having atomic numbers greater than 8 are very rare. For example, hydrogen — the smallest of all atoms — occurs with the highest 48.6% frequency. Next is carbon with a small atomic number of 6 which occurs with a relatively high 26.5% frequency. Next is nitrogen with the small atomic number of 7 which occurs with 9.0% frequency. Finally, it's oxygen with the small atomic number

ATOMIC NUMBER	ELEMENT SYMBOL	ELEMENT NAME	NUMBER IN CHOCOLATE	%
1	H	Hydrogen	26628277097	48.6%
6	C	Carbon	14532870466	26.5%
7	N	Nitrogen	4922146893	9.0%
8	O	Oxygen	8648578080	15.8%
9	F	Fluorine	149	<0.0001%
11	Na	Sodium	959308	0.0017%
12	Mg	Magnesium	21830440	0.0398%
15	P	Phosphorus	24860202	0.0453%
16	S	Sulfur	168	<0.0001%
17	Cl	Chlorine	86	<0.0001%
19	K	Potassium	41568596	0.0758%
20	Ca	Calcium	3360842	0.0061%
25	Mn	Manganese	76360	0.0001%
26	Fe	Iron	262489	0.0005%
27	Co	Cobalt	113	<0.0001%
30	Zn	Zinc	113072	0.0002%

Figure 1.7: Small Atoms Outnumber Big Atoms in Composition of Chocolate

of 8 which occurs with 15.8% frequency. All the other bigger atoms occur with a tiny frequency of no more than 0.08% each, and some are even less than 0.0001% frequency. Surely, such decisive preference for the small is expected to be found also in all other types of dishes, fruits, sweets, cookies, cakes, or whatever food. There is nothing special about chocolate that could give it any unique preference for small atomic constituents that other (standard) organic foods lack.

Note: Total estimated count of all the atoms in the chocolate piece is 54,824,904,359.

Hence percent occurrence of hydrogen atoms for example is calculated as the ratio:

$$(\# \text{ of hydrogen atoms})/(\text{total}) = (26{,}628{,}277{,}097)/(54{,}824{,}904{,}359)$$
$$= 0.486 = 48.6\%.$$

Chapter 5

Slow Spinning Pulsars Outnumber Fast Ones throughout the Universe

Pulsars are rotating neutron stars that emit a focused beam of electromagnetic radiation that is only visible if you're standing in its path. They are known as the 'lighthouses' of the universe. Pulsars aren't really stars, or at least they aren't living stars. Pulsars are formed when a massive star runs out of fuel in its core and collapses in on itself. This stellar death typically creates a massive explosion called a supernova. The neutron star is the dense nugget of material left over after this explosive death. Neutron stars are typically only about 20 to 24 kilometers in diameter, but they can contain up to twice the mass of the sun! They are the densest material in the universe, with the exception of whatever happens to matter inside a black hole. Pulsars are responsible for powerful gravitational forces in their neighborhood. Under such incredible pressure, matter behaves in ways not seen in any other environment in the universe.

At least 2560 pulsars have been detected in total since their first discovery in 1967. Most pulsars rotate on the order of once, twice, or thrice per second (these are called "slow pulsars"), while only about 200 pulsars rotate at the rate of hundreds of times per second (called "fast pulsars"). The fastest known pulsar rotates at the dizzying rate of 716.4 times per second! The hertz (symbol Hz) is the unit of frequency in the International System of Units (SI) and is defined as one cycle per second.

The Australian Telescope National Facility provides detailed and up-to-date data on all known pulsars. As of December 2, 2016, when the data was downloaded, 2560 known pulsars were shown.

The website is http://www.atnf.csiro.au/people/pulsar/psrcat/ (click on the 'Table' icon).

Examination of the data reveals that the distribution of spin frequency is highly skewed in favor of the slow. Figure 1.8 depicts the histogram of the pulsar data from 0 to 8 hertz. Clearly the slow (considered as small hertz) outnumbers the fast (considered as big hertz).

Out of 2560 hertz values in total, there are 2086 small hertz values from 0 to 8. There are 98 medium hertz values from 8 to 20. There are 376 big hertz values from 20 hertz to the maximum of 716.4 hertz, and these big values are sparsely spread, getting even more diluted and sparser as bigger hertz values are considered to the right of the histogram; hence the small is also beautiful for the tail of the histogram on the right (not shown in Figure 1.8). In other words, the histogram constructed exclusively for 20 to 720 hertz values would also fall to the right, favoring the "little big" over the "very big". For better visualization and economy of space only values up to 8 hertz are shown in the histogram of Figure 1.8.

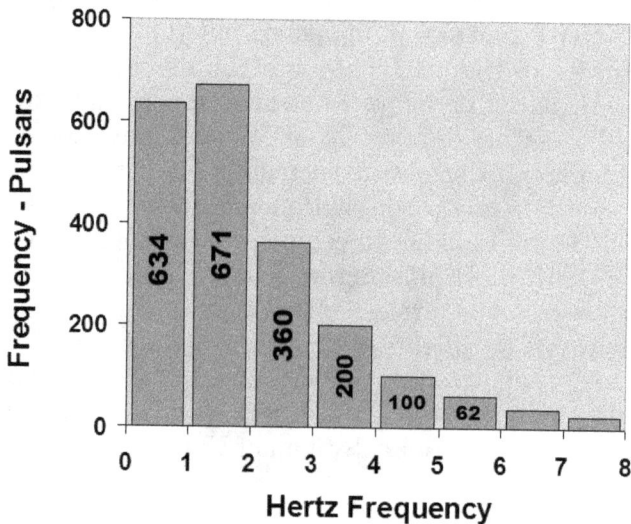

Figure 1.8:   Histogram of Pulsar Frequency — Slow Outnumbers Fast

There are	**634**	pulsars with frequency between	0 hertz	and	1 hertz.
There are	**671**	pulsars with frequency between	1 hertz	and	2 hertz.
There are	**360**	pulsars with frequency between	2 hertz	and	3 hertz.
There are	**200**	pulsars with frequency between	3 hertz	and	4 hertz.
There are	**100**	pulsars with frequency between	4 hertz	and	5 hertz.
There are	**62**	pulsars with frequency between	5 hertz	and	6 hertz.
There are	**36**	pulsars with frequency between	6 hertz	and	7 hertz.
There are	**23**	pulsars with frequency between	7 hertz	and	8 hertz.
There are	**17**	pulsars with frequency between	8 hertz	and	9 hertz.
There are	**23**	pulsars with frequency between	9 hertz	and	10 hertz.
There are	**11**	pulsars with frequency between	10 hertz	and	11 hertz.
There are	**8**	pulsars with frequency between	11 hertz	and	12 hertz.
There are	**7**	pulsars with frequency between	12 hertz	and	13 hertz.
There are	**4**	pulsars with frequency between	13 hertz	and	14 hertz.
There are	**6**	pulsars with frequency between	14 hertz	and	15 hertz.
There are	**9**	pulsars with frequency between	15 hertz	and	16 hertz.
There are	**6**	pulsars with frequency between	16 hertz	and	17 hertz.
There are	**2**	pulsars with frequency between	17 hertz	and	18 hertz.
There are	**0**	pulsars with frequency between	18 hertz	and	19 hertz.
There are	**5**	pulsars with frequency between	19 hertz	and	20 hertz.

Note: The author is relying on the collective wisdom and assumed honest work of the community of scientists providing this astronomical data and the theoretical framework regarding the existence of pulsars.

Chapter 6

Small Planets Outnumber Big Planets in the Milky Way Galaxy

Another instructive demonstration of the small is beautiful phenomenon in astronomy is found in data relating to all known exoplanets in our Milky Way Galaxy, namely planets within our home galaxy but outside the Solar System. The website http://exoplanet. eu/catalog/ provides data on those exoplanets. As of September 21, 2016, there were 1404 known exoplanets, each with an estimated mass value. This current count of 1404 exoplanets represents only a tiny fraction of the estimated 160 billion or so star-bound planets that exist in our home galaxy. Planets emit almost no light; they are dark objects out there, and therefore not easily detectable by us. Planets affect the orbit of their stars due to planet–star gravitational interactions; hence they could in principle be detected by observing the tiny wobble or deviation in the positions and motions of their light-omitting bright stars, but this requires delicate and more accurate astronomical observations.

The measurement of mass for the exoplanets is given in units of Jupiter, which is much heavier than planet Earth. The mass conversion formula (1 Jupiter) = (317.8 Earths) reminds us that Jupiter is truly massive in comparison with Earth. The other relevant mass conversion formula (1 Sun) = (1047.5 Jupiters) reminds us that Jupiter is truly tiny in comparison with its star.

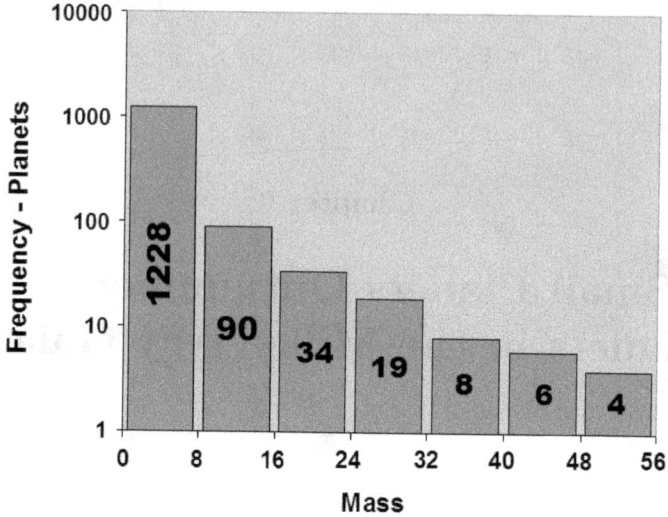

Figure 1.9:   The Small Decisively Outnumbers the Big in Exoplanet Mass Data

Figure 1.9 depicts the histogram of the mass of 1389 exoplanets having mass values from 0 to 56, in even steps of 8 for each successive bin — using the logarithmic scale for the vertical axis. There are 15 exceedingly massive and big exoplanets with mass over 56 Jupiters which are not shown in this histogram; they are excluded due to having mass values further to the right of the very last 48–56 bin. It should be carefully noted that logarithmic scales convert raw values into their logarithmic equivalents, and what should appear as 0, 1, 2, 3, 4 in Figure 1.9 is shown as 1, 10, 100, 1000, 10000 instead – as an alternative style.

The data on the mass of exoplanets upon which the above histogram was constructed can be summarized also as follows:

There are	**1228**	planets with mass between	0	and	8.
There are	**90**	planets with mass between	8	and	16.
There are	**34**	planets with mass between	16	and	24.
There are	**19**	planets with mass between	24	and	32.
There are	**8**	planets with mass between	32	and	40.
There are	**6**	planets with mass between	40	and	48.
There are	**4**	planets with mass between	48	and	56.
There are	**15**	planets having mass values over 56.			

Clearly, the small consistently and decisively outnumbers the big throughout this histogram.

If we focus only on the two extreme bins on the left-most and on the right-most sides of the histogram, we notice that there are 1228 (relatively) small exoplanets with mass less than 8 Jupiters, but only 4 (relatively) big exoplanets with mass between 48 and 56 Jupiters. Hence there are 307 times more very small exoplanets than very big ones!

The method of detecting the existence of an exoplanet is by way of the slight changes in the motion of its star, and such a method is biased against the small and in favor of the big. This is so since very small exoplanets cause the tiniest and most delicate changes in the motions of stars, and such subtle star movements are even harder for us to detect with the current technological level of equipment and instruments. Clearly, big exoplanets are much easier to detect than small exoplanets. Yet the small came up on top here in spite of its inherent disadvantage!

Because the small is beautiful effect is so dominant here, drawing the histogram with the original raw values for the vertical axis without taking the logarithm would reduce the big into such short and tiny bins on the right side that they would be very hard or simply impossible to visualize. The logarithmic scale for the vertical axis here allows better visualization for the data in its entirety and provides an alternative vista in understanding how relative quantities occur. Its drawback though is that it artificially reduces the visual disparity between the big and the small, although this is partially remedied by the numbers inscribed within the bins to reinforce the true nature of the quantitative configuration and the dramatic way the small outnumbers the big.

Figure 1.10 depicts the simple histogram based on the original raw data of exoplanets' mass, without using the logarithmic scale for the vertical axis. Visually this simple histogram is superior in a sense, as it depicts the true and dramatic disparity between the small and the big; yet, this comes at the heavy price of not allowing us to determine visually what is happening around big values on the right side — where everything falsely appeared reduced to zero height.

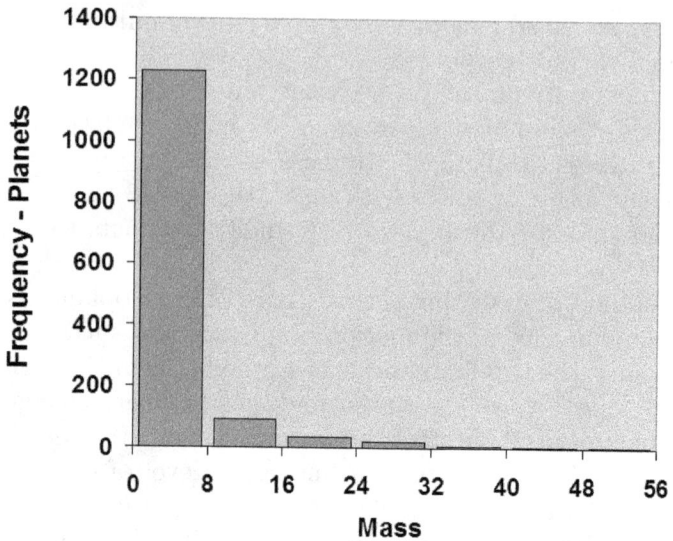

Figure 1.10: The Visual Difficulties in Drawing the Histogram of Exoplanet Raw Data

Figure 1.10 is presented here in order to provide one decisive example of the typical challenges and disadvantages in drawing histograms with the original raw values in many of the cases regarding the small is a beautiful phenomenon.

Chapter 7

Small Rivers Outnumber Big Rivers Worldwide

The list of the 181 most significant rivers worldwide, namely those that are longer than 1,000 kilometers, is available on wikipedia at: https://en.wikipedia.org/wiki/List_of_rivers_by_length.

The longest and most significant four rivers in the world are:

Amazon, 6992 kilometers long, spanning the countries of Brazil, Peru, Bolivia, Colombia, Ecuador, Venezuela, and Guyana. It's the longest river in the world!

Nile, 6853 kilometers long, spanning the countries of Ethiopia, Eritrea, Sudan, Uganda, Tanzania, Kenya, Rwanda, Burundi, Egypt, Congo, and South Sudan.

Yangtze, 6300 kilometers long, in China.

Mississippi, 6275 kilometers long, in the United States and Canada.

Figure 1.11 depicts the histogram of these 181 rivers. Here the raw or original scale is used for the vertical axis; since there is no compelling reason to use the logarithmic scale. The small is definitely more numerous than the big in worldwide rivers data.

The complete distribution is:

From 1000 to 2000 there are **119** rivers.
From 2000 to 3000 there are **35** rivers.
From 3000 to 4000 there are **13** rivers.
From 4000 to 5000 there are **7** rivers.
From 5000 to 6000 there are **3** rivers.
From 6000 to 7000 there are **4** rivers.

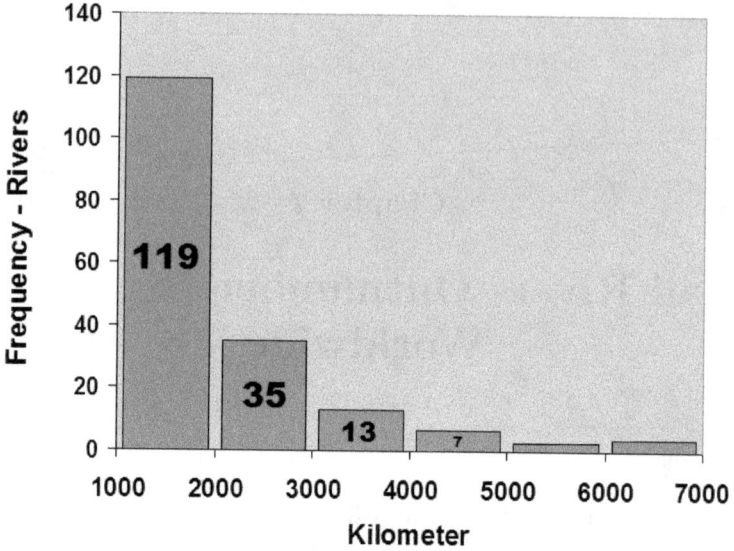

Figure 1.11: The Small Decisively Outnumbers the Big in Rivers Data

1006	1119	1236	1411	1610	2250	3060
1010	1120	1240	1415	1641	2270	3078
1010	1120	1242	1420	1670	2273	3180
1012	1123	1252	1420	1726	2287	3180
1015	1126	1270	1425	1749	2292	3185
1020	1130	1271	1438	1799	2333	3211
1030	1130	1280	1438	1800	2348	3380
1047	1130	1289	1465	1805	2410	3596
1049	1130	1300	1480	1809	2428	3645
1050	1130	1300	1490	1865	2450	3650
1050	1143	1302	1497	1870	2490	3672
1050	1143	1320	1500	1900	2500	4200
1067	1149	1320	1515	1927	2513	4241
1072	1150	1323	1532	1950	2549	4350
1078	1150	1345	1550	1978	2570	4400
1078	1158	1350	1575	2010	2615	4444
1080	1173	1352	1580	2092	2620	4700
1080	1175	1360	1591	2100	2620	4880
1086	1182	1364	1594	2100	2627	5410
1094	1190	1368	1599	2101	2650	5464
1100	1200	1370	1600	2102	2693	5539
1100	1210	1370	1600	2153	2888	6275
1100	1220	1372	1600	2170	2948	6300
1102	1223	1400	1600	2188	2989	6853
1105	1231	1400	1600	2200	3057	6992
1115	1233	1400	1600	2250	3058	

Figure 1.12: Data on Lengths of 181 Most Significant Rivers

Figure 1.12 depicts the entire data set on the lengths of these 181 most significant rivers in the world, sorted from low to high. This is shown for pedagogical purposes, to encourage the reader to attempt to make the intuitive or visual connection between how sorted small-is-beautiful-type data looks like numerically, and its innate skewed nature.

Chapter 8

Small Stars Outnumber Big Stars in the Universe

The small is beautiful phenomenon manifests itself decisively in the physical existence of planets and stars throughout the universe. The focus of this chapter is solely on the sizes of stars. As it happens, empirical astronomical data on stars strongly confirms the small is beautiful principle.

The layman, the non-expert, and readers who are less proficient in mathematics, physics, and astronomy, could simply skip this entire chapter on the sizes of stars and continue on to the next chapters without any loss of continuity, and without diminishing their ability to comprehend the entire small is beautiful phenomenon.

Let us exploit some well-established facts in astronomy regarding the typical types of stars that exist in the galaxies in order to learn about their relative sizes.

For centuries, astronomers wondered about the life cycles of stars. However, these questions proved difficult to answer because stars live so long. Even a relatively short-lived star burns for a few million years. Surely, a human with a lifespan of less than 100 years could never watch a star go through its complete life cycle; and even our entire civilization of about 4000 years is of too short a duration for such a study. Yet, by looking at a very large number of stars, we can see them in various stages of development. We see small stars, big stars, supergiant stars, young hot stars, old cool stars, and stars that have ended their lives and left empty shells. If at one (astronomical) instant we take a single comprehensive snapshot by looking at enough stars of various ages, sizes, temperatures, and luminosities, we can put together a complete picture of stellar evolution.

At the beginning of the 1900s, scientists closed in on a picture of stellar evolution. Physicists worked out the theory of nuclear fusion and realized that fusion provided enough energy to power stars. They realized that eventually, stars would run out of fuel for fusion and would burn out. So all stars would eventually die. But what would happen to stars during their lives?

The first clue came in 1911 when Ejnar Hertzsprung, a Danish astronomer studying at the University of Leiden in the Netherlands, plotted the luminosities of stars against their colors. Luminosity is a measure of the total energy a star gives off. The color indicates the star's surface temperature. In 1913, Henry Russell of Princeton University, independently plotted the same diagram that Hertzsprung made. The diagram became known as the Hertzsprung–Russell diagram, abbreviated HR diagram. Hertzsprung and Russell noticed some very particular and unexpected patterns in the diagram. By studying HR diagrams, later astronomers figured out the life cycles of stars. Figure 1.13 depicts the HR diagram, and this figure was originally adapted from the European Southern Observatory, CC-BY 4.0, and then further modified.

The Hertzsprung–Russell diagram is the most famous diagram in astronomy, and it is also referred to as the "atlas of the sky". It is a scatter plot of (optimally) all currently known stars in our Milky Way galaxy showing the relationship between the stars' luminosities versus their effective (surface) temperatures. The HR diagram uses the logarithmic scale for both variables, so that for each star, the diagram plots the logarithm of the star's brightness against the logarithm of its temperature. It has the unique and highly unconventional feature of having the horizontal x-axis increasing to the left, in sharp contrast to almost all other cartesian planes where the x-axis increases to the right.

The main discovery obtained in the plotting of the HR diagram is that almost all stars (~90%) fall within a long and relatively narrow band called the "Main Sequence", which stretches from the top-left part of the diagram, of hot, big, and bright stars, to the bottom-right part of the diagram, of cool, small, and dim stars. There are much fewer giant stars and supergiant stars clustered somewhere above the

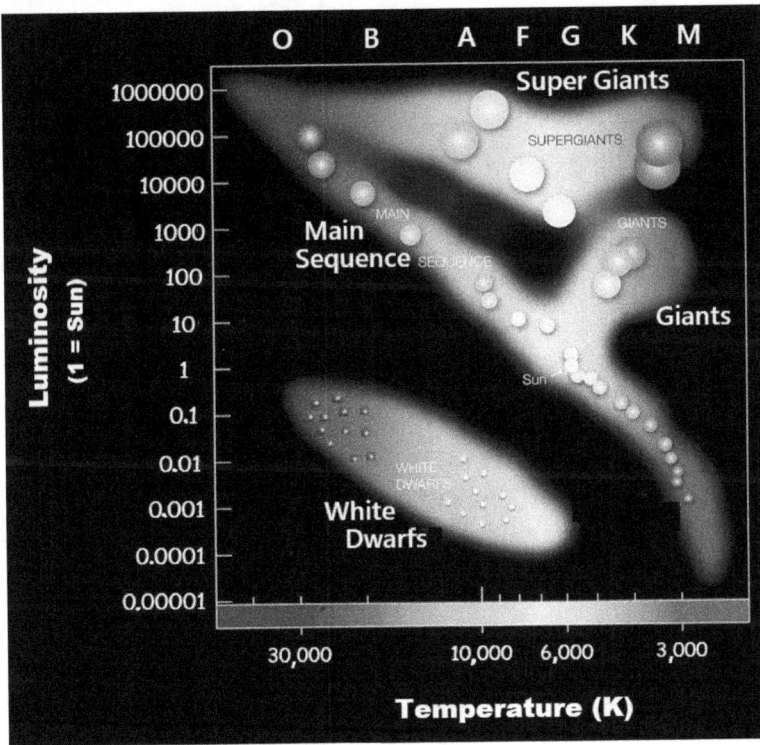

Figure 1.13: Hertzsprung–Russell Diagram — Luminosity vs. Temperature

Main Sequence; and there is another distinct cluster of much smaller stars called White Dwarfs which congregate below and to the left of the Main Sequence. Let us now examine some theoretical aspects of the diagram.

Stellar luminosity — namely the "brightness" of a given star — depends on its size, as well as on its surface temperature. The hotter the star, the brighter it shines. In addition, the bigger the star, the brighter it shines. The Stefan–Boltzmann law expresses these relationships as $L = 4\pi R^2 \sigma T^4$, where L is the luminosity, R is the stellar radius — which is a measure of its size and surface area, T is the temperature in kelvin, and alpha σ is the Stefan–Boltzmann constant. The law is only a very good approximation; nonetheless it enables astronomers to easily infer the radii of stars.

The three parallel inclined lines in the HR diagram of Figure 1.13 are each of a fixed star size, namely of a constant stellar radius R.

The higher lines are of bigger sizes, and the lower lines are of smaller sizes. To see why this is so, we simply take the logarithm of the Stefan–Boltzmann expression and obtain $\text{Log}(L) = \text{Log}(4\pi R^2 \sigma) + 4 * \text{Log}(T)$. For any fixed value of R, $\text{Log}(L)$ versus $\text{Log}(T)$ is linear with a slope of positive 4, and interpreted as a slope of negative 4 due to the inversion in the direction of the horizontal x-axis.

Note: $\log(x)$ or $\text{LOG}(X)$ notation in this book would always refer to our decimal base 10 number system, hence the more detailed notation of $\text{LOG}_{10}(X)$ is often — but not always — avoided. The natural logarithm base e would be referred to as $\text{LOG}_e(X)$ or simply $\ln(x)$.

Very bright but cool stars are therefore necessarily of very large sizes, since they do not possess sufficiently high temperature to shine much — if not for their unusually large sizes. These enormous stars are called red giants and lie above and to the right of the Main Sequence. A good example is the well-known star Antares, which is classified as a red supergiant. Antares is the fifteenth-brightest star in the night sky, and is likely among the largest of known stars. Its mass is calculated to be around 12 times that of the sun. Antares' estimated age is merely 12 million years, so the star is quite young, yet it is soon nearing the end of its life and is expected to explode violently as a supernova probably within the next few hundred thousand years. For a few months sometime in the future, the Antares supernova would be as bright as the full moon and be visible in daytime! Antares' temperature is 3548 K, which is quite "cool" compared with the sun's temperature of about 5770 K, yet Antares' luminosity is about 50,000 times brighter than the Sun! This must mean that Antares has a very large radius; and in fact it is about 600 times larger than the radius of the sun! Supergiant stars are suicidal, they are destined to end their lives violently; and they have short life spans of only between 30 million years and a few hundred thousand years. In sharp contrast, very dim but hot stars are necessarily of very small sizes, since they possess sufficiently high temperature to shine brightly — if not for their unusually small sizes. These small, dim, and hot stars are called white dwarfs. They could typically have radii as small as the Earth, and they have extremely high temperature of around 10,000 K.

Main Sequence stars are characterized by the source of their energy, which is to fuse hydrogen atoms to form helium atoms in their cores. These stars can range from about a tenth of the mass of the Sun up to 200 times as massive. Mass is the key factor in determining the lifespan of a Main Sequence star, its size and its luminosity. About 90% of the stars in the universe are Main Sequence stars (as is our sun); hence if this main group of stars exhibits the small is beautiful phenomenon, then we may conclude with confidence that such is the property of stars in general. The other types of stars which constitute merely 10% of all the stars in the universe are either red and blue giants, or white drafts, so this minority group of stars is kind of well-balanced between the big and the small, and thus does not sway the distribution of sizes by much.

To recap, regarding the small is beautiful phenomenon, everything hinges upon the size distribution of Main Sequence stars. Let us then examine this distribution.

The Harvard Spectral Classification allows us to divide stars into several spectral types depending on their temperature. This classification sequence is ordered from the hottest O-type stars to the coolest M-type stars via the designation O, B, A, F, G, K, M. This classification can be seen also on the top of the HR diagram in Figure 1.13. The following table in Figure 1.14 is obtained from https://en.wikipedia.org/wiki/Stellar_classification. The table summarizes the main spectral types in the Harvard Spectral Classification scheme. The units in use are: M for solar masses, R for solar radii, and L for bolometric magnitude of luminosity (radiant energy).

The last column on the right in the table plainly indicates that small stars of Class M with 76.5% fraction of all Main Sequence stars vastly outnumber big stars of Class O with only about

Class	Temperature	Main-Sequence Mass	Main-Sequence Radius	Main-Sequence Luminosity	Fraction of all M.S. Stars
O	$\geq$ 30,000 K	$\geq$ 16 M	$\geq$ 6.60 R	$\geq$ 30,000 L	0.00003%
B	10,000 – 30,000 K	2.10 – 16 M	1.80 – 6.60 R	25 – 30,000 L	0.1%
A	7,500 – 10,000 K	1.40 – 2.10 M	1.40 – 1.80 R	5 – 25 L	0.6%
F	6,000 – 7,500 K	1.04 – 1.40 M	1.15 – 1.40 R	1.5 – 5 L	3.0%
G	5,200 – 6,000 K	0.80 – 1.04 M	0.96 – 1.15 R	0.6 – 1.5 L	7.6%
K	3,700 – 5,200 K	0.45 – 0.80 M	0.70 – 0.96 R	0.08 – 0.6 L	12.1%
M	2,400 – 3,700 K	0.08 – 0.45 M	$\leq$ 0.70 R	$\leq$ 0.08 L	76.5%

Figure 1.14: Small Stars Vastly Outnumber Big Stars in Harvard Spectral Classification

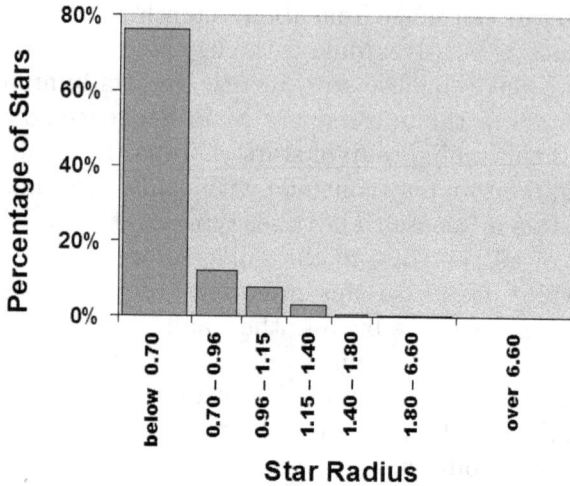

Figure 1.15:　Small Stars Decisively Outnumber Big Stars in the Universe

0.00003% fraction. Figure 1.15 depicts the histogram of the 7 stellar classes O, B, A, F, G, K, M, demonstrating that small stars decisively and consistently outnumber big stars. The horizontal scale is an arbitrary one and distorted a bit, not being drawn according to the true dimensions of the relevant sub-intervals; rather it is drawn according to the definition of the classes.

Disclaimer: The author is relying on the assumed honest work of the community of scientists providing this theoretical and empirical astronomical analysis. Since science also involves political, financial, and other social factors which at times might adversely affect its development, a bit of healthy skepticism and further scrutiny is warranted regarding scientific publications and paradigms. To this author, subjectively, it seems that in the past 150 or so years, science had unnecessarily overreached itself with grand conjectures, guesses, and speculations, built upon even more grandeur earlier conjectures, guesses, and speculations. Hence the author is asking for forgiveness and understanding from the readers if some of the scientific assertions quoted in this chapter shall prove to be false or without sufficient proofs and supporting evidence.

Chapter 9

Small Cities and Towns Outnumber Big Metropolises

Data on the USA 2009 Population Census of all incorporated cities and towns shall be analyzed in terms of its size configuration. Detailed data on the USA Population Census can be obtained via the link: https://census.gov/data/tables.

This data set on the populations of all 19,509 incorporated cities and towns starts at value 1, namely a single person living in an officially recognized town. Its top value is that of New York City with a population of 8,391,881.

Figure 1.16 depicts the histogram from population of 1 to population of 175,000, in even steps of 25,000 for each successive bin. The notation "k" denotes 1,000 people. The vertical axis uses the logarithmic scale for better visualization, although this masks the dramatic fall in the histogram. Because the histogram is limited to those cities below 175,000 inhabitants, it contains only 19,377 cities and towns, excluding 132 very big cities over 175,000.

The small is decisively more numerous than the big in this population data.

There are numerous tiny incorporated towns and very small cities that exist almost anonymously, and their names, history, or unique architecture are not even known to most people, except for the local inhabitants. On the other hand, the nine largest cities in the

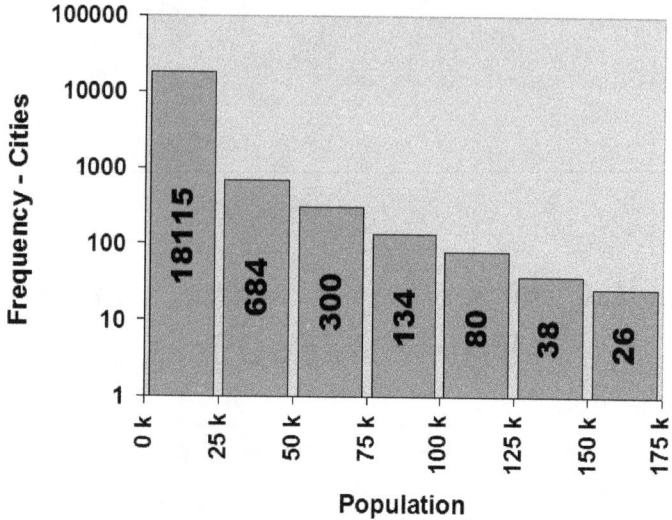

Figure 1.16: The Small Decisively Outnumbers the Big in USA Population Data

USA with over one million inhabitants are well-known and rather famous:

New York City, New York	8,391,881
Los Angeles, California	3,831,868
Chicago, Illinois	2,851,268
Houston, Texas	2,257,926
Philadelphia, Pennsylvania	1,593,659
Phoenix, Arizona	1,547,297
San Antonio, Texas	1,373,668
San Diego, California	1,306,300
Dallas, Texas	1,299,542

Even within this very short list of the biggest nine cities, one could easily detect the small is beautiful phenomenon! There are **5** cities between one million and two million; there are **2** cities between two million and three million; but there is only **1** city between three million and four million!

Chapter 10

Small Wars Outnumber Big Wars in History

Wikipedia provides data on some 134 wars listed by death toll. Death estimates are of all types, either directly in battle of military personnel, or indirectly such as deaths of civilians, epidemics, diseases, famines, and genocides caused by war. The link to the page is at: https://en.wikipedia.org/wiki/List_of_wars_by_death_toll.

WWII (1939–1945) tops the list, with 58,309,519 total deaths. Surprisingly, WWII is not an outlier, an anomaly, or extraordinarily bloody, as it is closely followed by the Mongol Conquests (1206–1324) forged by Genghis Khan, with 52,915,026 total deaths, and according to some scientists and historians it was spreading the bubonic plague (also known as the Black Death) across much of Asia and Europe, and causing massive loss of life. This is followed by the third deadliest but little known war of the Taiping Rebellion in China (1850–1864) with 44,721,360 total deaths.

Some noteworthy wars in the list:

The Napoleonic Wars in Europe (1803–1815) — 4,600,000 deaths.
The American Civil War (1861–1865) — 800,000 deaths.
The Mexican Revolution (1910–1920) — 1,414,214 deaths.
World War I (1914–1918) — 20,000,000 deaths.
The Spanish Civil War (1936–1939) — 500,000 deaths.
The Korean War (1950–1953) — 1,200,000 deaths.
The Vietnam War (1965–1975) — 1,743,560 deaths.

The horrific number of 487,767,309 estimated total deaths from all these wars — especially in comparison to the much smaller global population in past eras — is a moving testimony to the hidden dark forces lurking underneath us at all times, casting a shadow over our relatively peaceful existence of recent years. The author wishes to express his strong sense of revulsion and rejection of this appalling long list of wars in recorded history, hoping for a better humanity.

Surely the list is not as accurate as it could be, and surely there are many ancient wars missing from the records, but the phenomenon of observing numerous smaller entities but only few big ones can still be clearly deciphered here as well. In the context of wars, the small is beautiful phenomenon is definitely desired and appreciated! Luckily for us, the biggest and bloodiest wars were relatively very few. Most wars were smaller, with much lower death toll, and this fact provides us with some limited solace.

Figure 1.17 depicts the histogram of the frequency of occurrences of wars by number of deaths. The horizontal x-scale is an arbitrary one, chosen in order to obtain a good visual presentation of this very small data set. This is almost a logarithmic scale, except for that arbitrary jump or gap on the left from 0 to 100,000. The bin

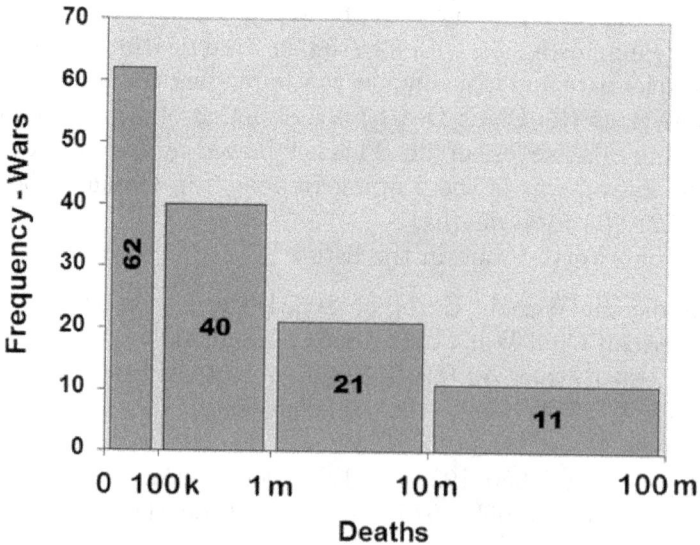

Figure 1.17:   Many Small Wars but Only Few Big and Bloody Wars in History

on the left-most side of the histogram (of the smallest values) is given the narrowest interval of only 100,000. The next bin is with an interval of 900,000, followed by the next bin with an interval of 9,000,000. The bin on the right-most side of the histogram (of the biggest values) is with the widest interval of 90,000,000. Yet, in spite of this strong advantage for the relatively big over the relatively small in consistently obtaining wider intervals throughout the range, big wars could not prevail over small wars.

Chapter 11

The Poor Vastly Outnumber the Rich in Global Wealth Distribution

Wikipedia provides an estimate of wealth distribution worldwide. The link is at: https://en.wikipedia.org/wiki/Distribution_of_wealth.

Assuming a global population of 7 billion, wealth levels (in US dollars) are distributed as follows:

4,809,000,000 people with wealth less than $10,000 (68.7% of global population).

1,603,000,000 people with wealth between $10,000 and $100,000 (22.9% of global population).

539,000,000 people with wealth between $100,000 and $1,000,000 (7.7% of global population).

49,000,000 people with wealth over $1,000,000 (0.7% of global population).

Figure 1.18 depicts the histogram of the number of people by wealth. The vertical scale is in units of billion. The horizontal x-axis scale is an arbitrary one, chosen in order to obtain a good visual presentation of the data. In spite of the strong advantage for the relatively big over the relatively small in consistently obtaining wider intervals throughout the range, the big/rich is vastly outnumbered by the small/poor.

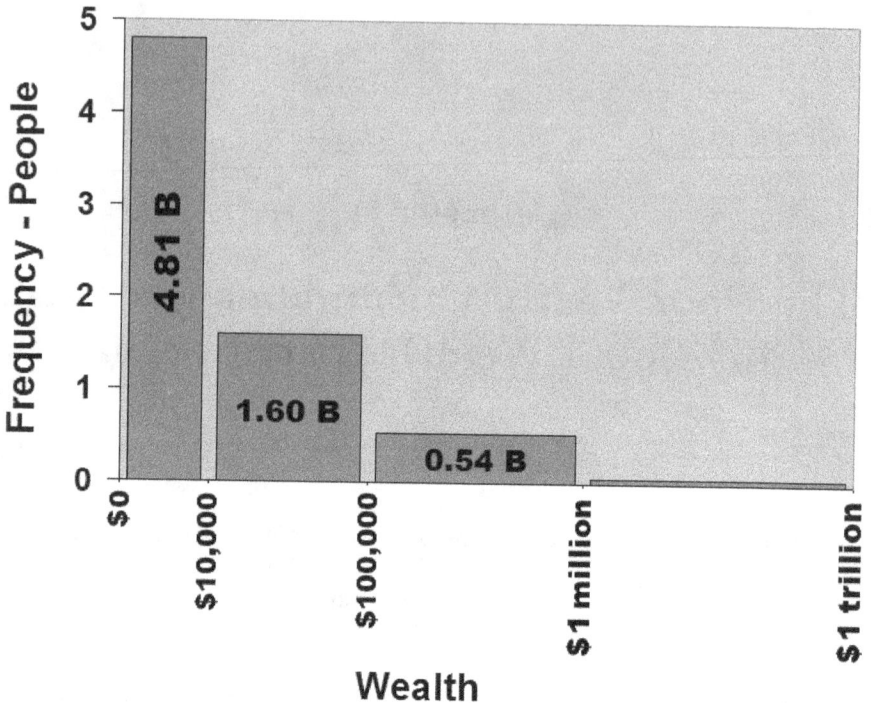

Figure 1.18: The Poor Decisively Outnumber the Rich — Global Wealth Distribution

Disclaimer: The author is not attempting to endorse or justify such an unhappy state of economics affairs, nor to rebel against and protest such terrible inequality experienced by humanity. This chapter on wealth distribution is only meant to account for the statistical and factual economic and financial situation occurring throughout the world, without referring to any ethical, moral, or political points of view.

Chapter 12

Small Companies Outnumber Big Corporations

The list of 2889 companies registered on the NASDAQ Exchange — together with market capitalization values as of end of October 9, 2016 — was downloaded from the NASDAQ website at: http://www. nasdaq.com/screening/companies-by-industry.aspx.

Market capitalization is defined as (Current Stock Price) × (Shares Outstanding), hence it can be thought of as the "price" of the entire company. Surely if a single superwealthy investor (such as, for instance, Warren Buffett) wishes to become the sole owner of the corporation, then that person would need to purchase all the outstanding shares at the current market price from all the numerous smaller investors (assuming all are willing to sell). Market capitalization signifies the size of the company for the most part, and it correlates fairly well with other measurements of company size, such as number of employees, revenues, expenses, tax due, or profits.

Figure 1.19 depicts the histogram of market capitalization from 0 to 2 billion dollars, in even steps of $250,000,000 for each successive bin. In a consistent and monotonic manner, the smaller companies outnumber the bigger companies on the entire range.

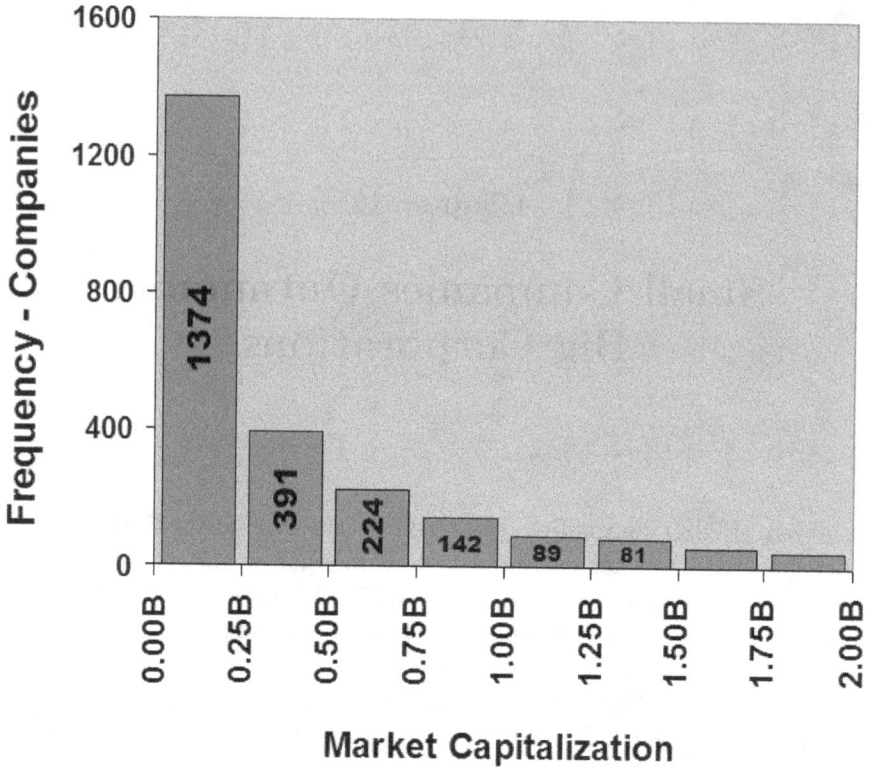

Figure 1.19:   Small Companies Decisively Outnumber Big Companies

The summary table of the number of companies within $0 to $3,000,000,000 is as follows:

From	Up To	Companies
0	250,000,000	1374
250,000,000	500,000,000	391
500,000,000	750,000,000	224
750,000,000	1,000,000,000	142
1,000,000,000	1,250,000,000	89
1,250,000,000	1,500,000,000	81
1,500,000,000	1,750,000,000	59
1,750,000,000	2,000,000,000	48
2,000,000,000	2,250,000,000	31
2,250,000,000	2,500,000,000	42
2,500,000,000	2,750,000,000	27
2,750,000,000	3,000,000,000	23

There are 481 super big companies with market capitalization over \$2 billion which are not shown in the histogram of Figure 1.19, being very thinly spread out further to the right, and the tail of the histogram keeps falling there as well. The biggest five companies are:

Facebook Inc. with \$368,920,000,000 in market capitalization (\$368.92B).

Amazon.com, Inc. with \$400,290,000,000 in market capitalization (\$400.29B).

Microsoft Corporation with \$449,160,000,000 in market capitalization (\$449.16B).

Google Inc. GOOG with \$533,650,000,000 in market capitalization (\$533.65B).

Google Inc. GOOGL with \$550,660,000,000 in market capitalization (\$550.66B).

Apple Inc. with \$609,160,000,000 in market capitalization (\$609.16B).

Chapter 13

Small Bills Outnumber Big Bills in Revenue and Expense Data

Almost all accounting, financial, and economics data sets are structured in such a way that the small is more numerous than the big, with very few and very rare exceptions. This fact can be easily verified by examining such data; although obtaining actual revenue and expense accounting data for any particular company or corporation is nearly impossible due to confidentiality, privacy, and secrecy issues. Only quarterly or annual financial statements are public information that can be readily examined, but the small is beautiful phenomenon is encountered mostly in the raw detailed values (at the individual bill, receipt, and invoice level), and not really in summary values, aggregations, and ratios of values, such as the numbers typically found in the financial statements.

Fortunately for statisticians, some governmental expense data is available at the raw and original level, detailing each and every transaction.

The State of Oklahoma in the USA provides detailed information at the transaction level for all its vendor payments for the fiscal year 2011. The website for this database is at: https://data.ok.gov/dataset/state-oklahoma-vendor-payments-fiscal-year-2011.

These payments reflect disbursements from a state fund for the purchase of goods received, services performed, reimbursements, and payments to other governments.

Although this data set is purely of expenses and costs, it also reflects strongly on revenue data in general. This is so since every entry here for a given expense is surely also one revenue item for some provider, company, or agent, billing the state and charging it for the product or service rendered.

Examination of this data set reveals that the distribution of expense amounts is highly skewed in favor of the small. Figure 1.20 depicts the histogram of the vast majority of expenses from $0 to $1 million, containing 986,962 items, leaving out only a small minority of 530 very expensive items of over $1 million. The horizontal x-axis scale is mostly a logarithmic one, except for that jump or gap from $0 to $100. Clearly, except for a brief and very gentle rise on the left of the histogram between $0 and $1000, the cheap (considered as small) outnumbers the expensive (considered as big). The temporary and minor reversal of the phenomenon in the beginning for very low values is quite typical in revenue and expense accounting data, yet the overall description of relative quantities is decisively in favor of the small.

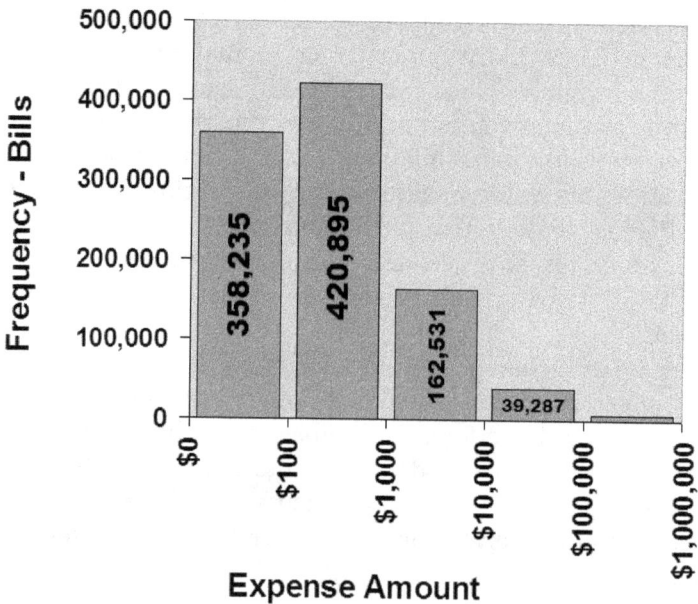

Figure 1.20: Oklahoma State Expenses — Cheap Outnumbers Expensive

The complete distribution is:

There are	**358,235**	invoices from	$0	up to	$100.
There are	**420,895**	invoices from	$100	up to	$1,000.
There are	**162,531**	invoices from	$1,000	up to	$10,000.
There are	**39,287**	invoices from	$10,000	up to	$100,000.
There are	**6,014**	invoices from	$100,000	up to	$1,000,000.
There are	**523**	invoices from	$1,000,000	up to	$10,000,000.
There are	**7**	invoices from	$10,000,000	up to	$100,000,000.

Note: Because this data set regarding Oklahoma State expenses is highly detailed and comprehensive, not pertaining to sums or aggregations, it is truly huge, containing too many records. Due to computer capacity and speed limitation, only the first 1,000,000 records from the top were selected for analysis, and the rest of the bills ignored. Out of these one million records, 12,508 bills with negative values were also ignored, leaving only 987,492 zero or positive-valued bills for quantitative analysis. The minimum positive bill is $0.01, namely one cent. The maximum bill is $99,115,421.

Chapter 14

Cheap Items Outnumber Expensive Items in Catalogs and Price Lists

Except for very small shops or highly specialized retail stores narrowly focused on some very particular items with little variety in the set of products on sale, the typical catalogs and price lists of almost all medium to large stores, shops, and retail establishments, show a definite preference for the cheap (small values) against the expensive (big values). In other words, there are numerous cheap items for sale; there are some medium-priced items for sale; but there are only very few very expensive items for sale.

In order to provide an example of the typical quantitative structure of large catalogs, the price list of Canford Audio PLC is selected and quantitatively analyzed. This company retails and manufactures a wide variety of electronic items aimed at those who use or install audio, video, and communications equipment, including the whole spectrum of data hardware and infrastructure products. There are 14,914 items for sale listed on their website.

The page of the price list on its website is at: https://www. canford.co.uk/PriceList/ and US dollars (\$) is selected as the currency. The prices quoted in column H are chosen, pertaining to the purchase of a single quantity, ignoring the discounted prices for the purchase of multiple quantities.

The cheapest item on sale is "JST CONNECTOR Crimp contact", offered at \$0.03, or 3 cents.

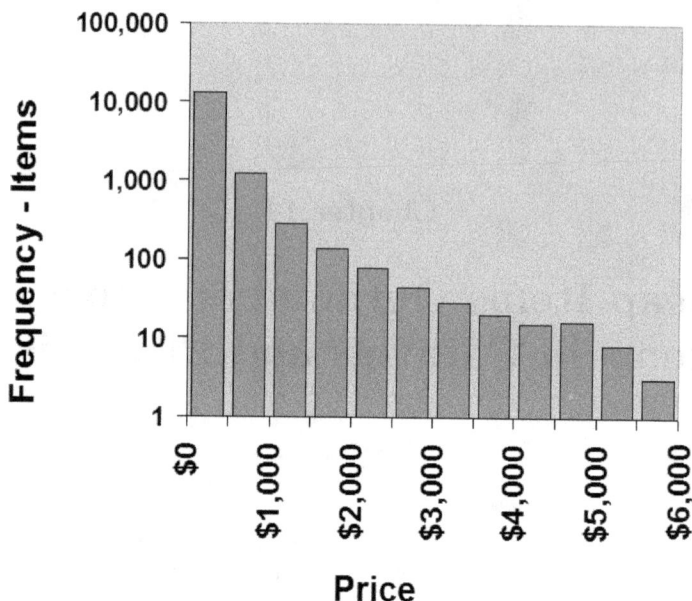

Figure 1.21:   Canford Audio Price List — Cheap Outnumbers Expensive

The most expensive item on sale is "NEUTRIK NOCS-WL-8x4-1310 OPTICAMSWITCH Wieland/LC FO connection", offered at $17,015.10.

Figure 1.21 depicts the histogram of almost all the items in Canford's price list, namely 14,885 items costing less than $6,000 — missing out only 29 very expensive items costing from $6,000 to the maximum of $17,015. The vertical scale of the histogram is constructed using the logarithmic scale. Such a scale visually suppresses the enormous preference for the small here which is indeed much more dramatic than what is seen in the histogram.

Needless to say, the small is beautiful phenomenon found in price lists and catalogs often (but not always) constitutes the cause and the driving force behind the manifestation of the same phenomenon in revenue and expense accounting data. The reason for this obvious connection is that an expense or revenue item (i.e. invoice or bill for the purchase of several items) is nothing but the combination of prices from the price list of a provider, often with repeated amounts if the shopper buys several quantities of a desired item.

The complete distribution up to $6,000 is:

There are	13052	items for sale with price between	$0	and	$500.
There are	1205	items for sale with price between	$500	and	$1,000.
There are	280	items for sale with price between	$1,000	and	$1,500.
There are	135	items for sale with price between	$1,500	and	$2,000.
There are	78	items for sale with price between	$2,000	and	$2,500.
There are	44	items for sale with price between	$2,500	and	$3,000.
There are	29	items for sale with price between	$3,000	and	$3,500.
There are	20	items for sale with price between	$3,500	and	$4,000.
There are	15	items for sale with price between	$4,000	and	$4,500.
There are	16	items for sale with price between	$4,500	and	$5,000.
There are	8	items for sale with price between	$5,000	and	$5,500.
There are	3	items for sale with price between	$5,500	and	$6,000.

Chapter 15

Small House Numbers Outnumber Big House Numbers in Address Data

The small is beautiful phenomenon is found consistently in address data, provided that the relevant data set is large enough. Here the focus is not on the street number or the zip code in any way. The focus is exclusively on the house number.

Address data pertaining to Prince Edward Island in Eastern Canada can be downloaded at the link in the website: http://www.gov.pe.ca/civicaddress/download/.

In total, there are 23,633 addresses on the island. Examination of the data reveals that there are many houses with small numbers but only few houses with big numbers. Figure 1.22 depicts the histogram of 17,422 house numbers in the address data from house number 1 to house number 1000. There are also 6,211 house numbers beyond 1000 which are not shown in the histogram, and which are thinly spread out and continue to fall further to the right of the histogram.

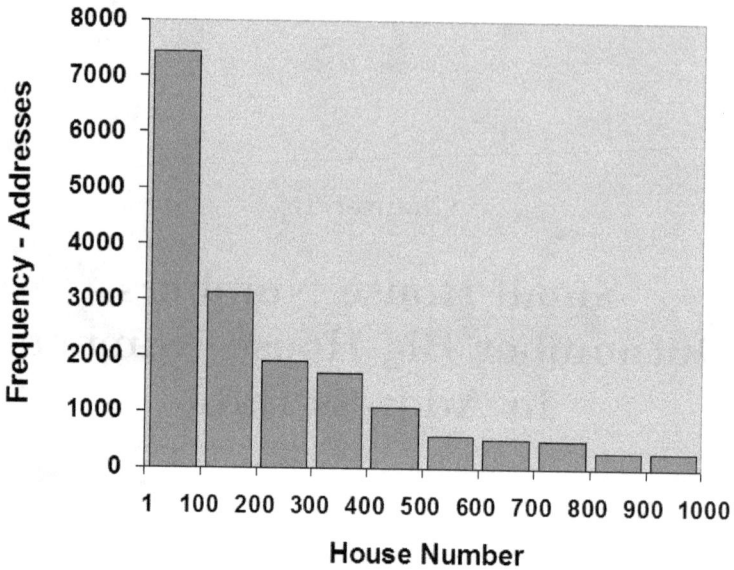

Figure 1.22:   Many Houses with Small Numbers but Only A Few Houses with Big Numbers

A small sample of ten selected addresses, including the smallest house number of 1, as well as the biggest house number of 43015, is outlined below:

1	Chestnut Street
1	Wellington Road
9	Pine Drive
15	Duck Pond Lane
27	Brooks Avenue
95	Kelly Drive
121	Beaton Rd — Route 138
318	Beaton Rd — Route 138
2257	Beaton Rd — Route 138
43015	Western Rd — Route 2

Chapter 16

Global Carbon Dioxide Pollution Data in Favor of the Small

Data pertaining to CO_2 (carbon dioxide) emissions by 216 sovereign states and territories in 2008 shall be analyze quantitatively. The data is downloaded from: http://en.wikipedia.org/wiki/List_of_countries_by_carbon_dioxide_emissions.

Each data point represents annual emission in thousands of metric tons for a given country.

The world's top six polluters are: China, 7,031,916; United States, 5,461,014; European Union, 4,177,817; India, 1,742,698; Russia, 1,708,653; and Japan, 1,208,163.

In order to draw a concise histogram for this data set with only a few bins, a reasonable criterion for definitions of four sizes shall be established as follows:

1st bin: from 0 to 10,000, signifying the **very small**. Range here is 10,000.

2nd bin: from 10,000 to 100,000, signifying the **somewhat small**. Range here is 90,000.

3rd bin: from 100,000 to 1,000,000 signifying the **medium**. Range here is 900,000.

4th bin: from 1,000,000 to 10,000,000 signifying the **big**. Range here is 9,000,000.

Counting occurrences of sizes according to this criterion yields the following result:

There are **117** countries with **very small** pollution levels between 0 and 10,000.

There are **60** countries with **somewhat small** pollution levels between 10,000 to 100,000.

There are **33** countries with **medium** pollution levels between 100,000 to 1,000,000.

There are **6** countries with **big** pollution levels between 1,000,000 and 10,000,000.

Figure 1.23 depicts the histogram. Clearly the small decisively outnumbers the big in the carbon dioxide data set, and this is so in spite of giving medium and big better chances of occurring by allocating them much wider ranges than the range allocated for small. For better visualization the histogram is not drawn exactly according to these four distinct scales for the ranges. In other words, true proportions between ranges vary sharply, yet they appear

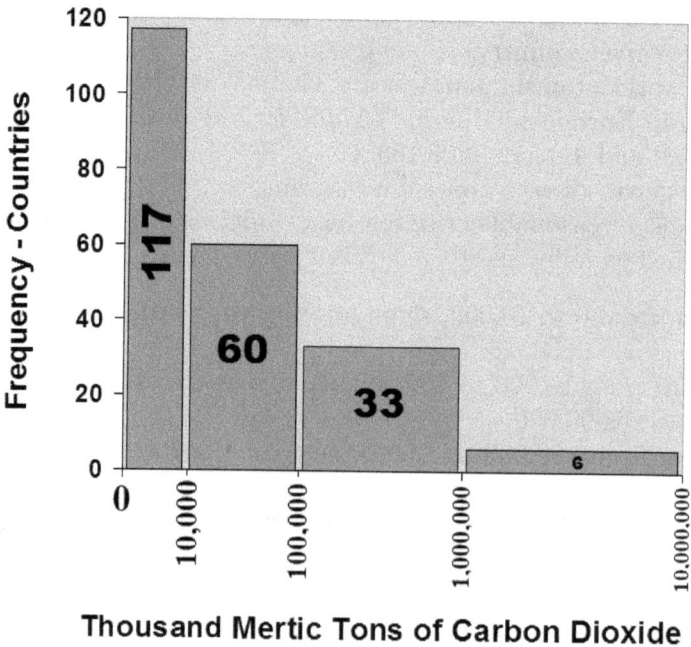

Figure 1.23:   The Small Outnumbers the Big in Carbon Dioxide Data

unrealistically mild and almost equal in the chart. For example, the range 1,000,000–10,000,000 is tenfold wider than the range 100,000–1,000,000, yet it appears only slightly wider in the histogram for scarcity of space and better visualization. The horizontal x-axis scale is almost a logarithmic one, except for that arbitrary big jump or gap on the left from 0 to 10,000.

Chapter 17

Small Families Outnumber Big Families Regarding Number of Children

Family size adheres to the small is beautiful principle in the modern era for most countries and societies in Europe, Asia, and the northern English-speaking part of North America; except for most of Africa, the Middle East, and some parts of Latin America, where the birth rate is relatively high. In Europe, Asia, and Northern America, there are many more small families than big families; and by far the most common family size is 1 to 2 children; while big families with 3, 4, 5, or more kids are mostly rare. The United Kingdom Office for National Statistics provides data for "Families with dependent children by number of children" for the year 2015. The link to their website can be found at: https://www.ons.gov.uk/peoplepopulationandcommu nity/birthsdeathsandmarriages/families/adhocs/006154familieswith dependentchildrenbynumberofchildrenuk1996to2015.

Out of a total of 7,927,000 families surveyed, 3,590,000 families were with a single child; 3,171,000 families were with 2 children; 883,000 families were with 3 children; 210,000 families were with 4 children; and only 73,000 families were with 5 or more children. Figure 1.24 depicts the family size histogram, visually confirming the fact that small families are much more numerous than big families.

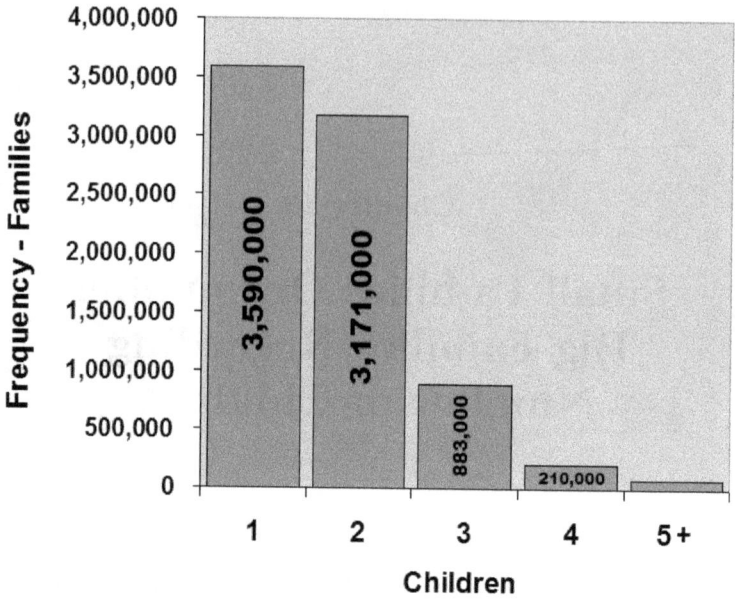

Figure 1.24: The Small Outnumbers the Big in UK Family Size Data

Chapter 18

Small Meteorites Outnumber Big Meteorites

Meteorites are rocks that originate in outer space and fall to the surfaces of planets and their satellites, such as our planet Earth and its moon. When the original object (termed "meteoroid") enters the atmosphere, various factors such as friction, pressure, and chemical interactions with the atmosphere cause it to heat up, and consequently it becomes a meteor and forms a fireball. When *meteoroids* enter Earth's atmosphere, or that of another planet or satellite, at high speed and burn up, the fireballs or "shooting stars" are called *meteors*. When meteoroids survive their trip through the atmosphere and hit the ground, they are called *meteorites*. Meteorites vary greatly in size, and some meteorites are large enough to create an impact crater on the ground they fall into, as noted and recorded by geologists. Meteorites are divided into three broad categories: (i) stony meteorites that are rocks, mainly composed of silicate minerals; (ii) iron meteorites that are largely composed of ferronickel; and (iii) stony-iron meteorites that contain large amounts of combination of metallic and rocky materials.

The website <https://catalog.data.gov/dataset/meteorite-landings> provides extensive meteorite data from the Meteoritical Society, and harvested from NASA sources. It contains information on all of the known 45,566 meteorite landings worldwide in all recorded history, and this metadata was last updated in December 2023. The variable in focus here is the column on mass in weight of grams, signifying the size. The smallest meteorite is of 0.01 gram.

The meteorite called Hoba is the largest known intact meteorite on Earth's surface. It fell in prehistoric times in Namibia, southwest Africa. The main mass of Hoba is estimated to be about 60,000,000 grams, and which is more than 60 tones. The Hoba meteorite was found in 1920, and it is thought to have impacted Earth about 80,000 years ago.

Figure 1.25 depicts the histogram of 39,153 meteorites out of the total 45,566 meteorites in the data set of the above website. The histogram is constructed in even steps of 60 grams for each successive bin. The histogram incorporates values from 0 gram to 600 grams, while missing only 6,413 very big meteorites of over 600 grams, which are thinly spread out, getting more diluted and rare as we move along the x-axis in the positive direction of the gram weight. The vertical axis uses the logarithmic scale for better visualization, although this masks the sharp fall in the histogram, as it visually suppresses the enormous preference for the small here which is indeed much more acute and dramatic as compared to what is seen in

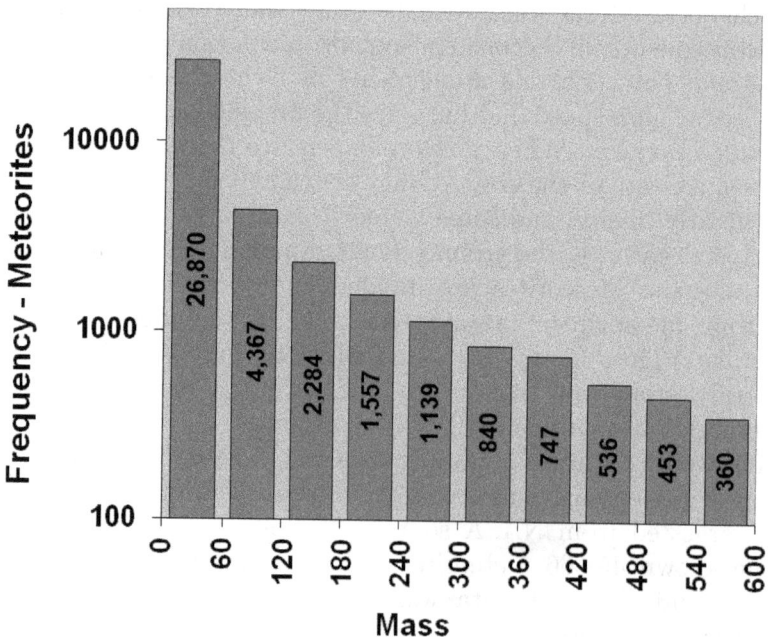

Figure 1.25: Small Meteorites Outnumber Big Meteorites Falling to Earth

the histogram of Figure 1.25. In conclusion: small-size meteorites decisively outnumber big-size meteorites, and this fact is cherished and deemed as a blessing and as essential by all the inhabitants and creatures of this planet, animals, and humans alike, who favor safety, and do not wish to have big rocks constantly falling over their heads.

Chapter 19

Small Volcano Eruptions Outnumber Big Volcano Eruptions

On the webpage <https://disasters.amerigeoss.org/pages/volcano> and under the link <https://disasters.amerigeoss.org/datasets/3ed 5925b69db4374aec43a054b444214_6/about> data on the history of past volcano eruptions can be found below the title "Historical Significant Volcanic Eruption Locations", which allows users to download detailed data and information about each significant eruption.

A significant eruption is classified as one that meets at least one of the following criteria: (i) caused fatalities, (ii) caused moderate damage of approximately $1 million or more, (iii) Volcanic Explosivity Index or VEI of 6 or greater, (iv) generated a tsunami, or (v) was associated with a significant earthquake.

The database contains information on the latitude, longitude, elevation, type of volcano, last known eruption, VEI, and socio-economic data such as the total number of casualties, injuries, houses destroyed, houses damaged, and dollar damage estimates. The Significant Volcanic Eruptions Database is a global listing of over 800 eruptions from 4360 BC to the present 2024 AD, namely data covering six millennia and some.

An arbitrary yet reasonable definition of sizes is obtained by focusing exclusively on the number of fatalities in each eruption. Hence according to this measure, a big eruption implies many fatalities, while a small eruption implies only few fatalities.

The data set contains a total of 877 significant volcano eruptions; 437 of those events are without specifically recorded number of fatality values and which will be excluded from our analysis; leaving us focused on only 440 events which are with an explicit estimate of non-zero number of fatalities; from the more gentle ones with just 1 fatality, to the most horrific one regarding the major plinian eruption associated with caldera formation, and which destroyed the early Mayan cities of Central America in 450 A.D. (plus or minus 30 years); with estimated 30,000 fatalities in the area, mostly those unfortunate victims who were swept by the pyroclastic flows.

A pyroclastic flow is a hot (typically over 800°C), chaotic mixture of rock fragments, gas, and ash that travels rapidly, at around tens of meters per second, away from a volcanic vent or collapsing flow front. Pyroclastic flows can be extremely destructive and deadly because of their high temperature and mobility.

Figure 1.26 depicts the histogram of 383 significant volcano eruptions having between 1 and 239 fatalities, out of the total of 440 eruptions with known fatality figures in the data set, thus missing only 57 very big eruptions of 240 or more fatalities which are thinly

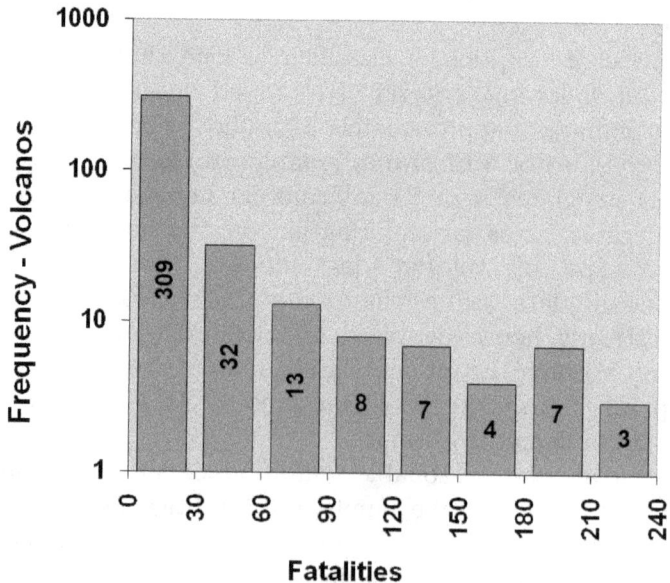

Figure 1.26: Small Volcanoes Decisively Outnumber Big Volcanoes

spread out, getting more diluted and rare as we move along the x-axis in the positive direction representing fatalities. The histogram conveniently starts at 0 actually, and it is constructed in even steps of 30 fatalities for each successive bin. The first bin incorporates all eruptions with fatalities from 0 to 30, but not including 30, namely the interval $[0, 30)$, although only non-zero eruptions are considered here beginning at 1, as we excludes the records of eruptions without unspecified number of fatalities. The second bin incorporates all eruptions with fatalities from 30 to 60 but not including 60, namely the interval $[30, 60)$, then $[60, 90)$, $[90, 120)$, and so on to $[210, 240)$. The vertical axis uses the logarithmic scale for better visualization, although this masks the sharp fall in the histogram, as it visually suppresses the enormous preference for the small here which is indeed much more acute and dramatic as compared to what is seen in the histogram of Figure 1.26. In conclusion: small-size volcano eruptions decisively outnumber big-size volcano eruptions, and this fact is welcome, cherished, and deemed as a blessing and as essential by all the inhabitants living around the edges and at the foothills of these menacing and powerful mountains containing hot and harmful lava.

Chapter 20

Small Countries Outnumber Big Countries

Wikipedia provides data on the areas of all countries and dependencies worldwide on the page: <https://en.wikipedia.org/wiki/List_of_countries_and_dependencies_by_area>.

The unit of square kilometers shall be chosen for our purposes. Areas for 253 countries and dependencies are listed there.

The biggest country in the world is Russia, with its rich culture and unique history, well-known for talented chess players, ballet dancers, and rich literary tradition with renowned writers such as Leo Tolstoy, Fyodor Dostoevsky, and Anton Chekhov. Perhaps its immense size explains the numerous highly acclaimed and revered mathematicians and scientists with which Russia has endowed the world, the most famous of them being Dmitri Mendeleev, who formulated the Periodic Law in 1869, a discovery that is viewed as the most significant contribution to materials chemistry, leading directly to the Periodic Table. Other personalities are Mikhail Lomonosov, regarded as the first to discover the law of mass conservation in chemistry in 1760; Nikolay Lobachevsky, the founder of hyperbolic geometry in 1829, which was later recognized as a valid alternative to the Euclidean geometry of the ancient Greeks; and closer to heart for this writer, condemned by fate to deal always with endless and demanding statistical matters, is Pafnuty Chebyshev, who made several breakthrough discoveries in mechanics, mathematics, and especially in statistics.

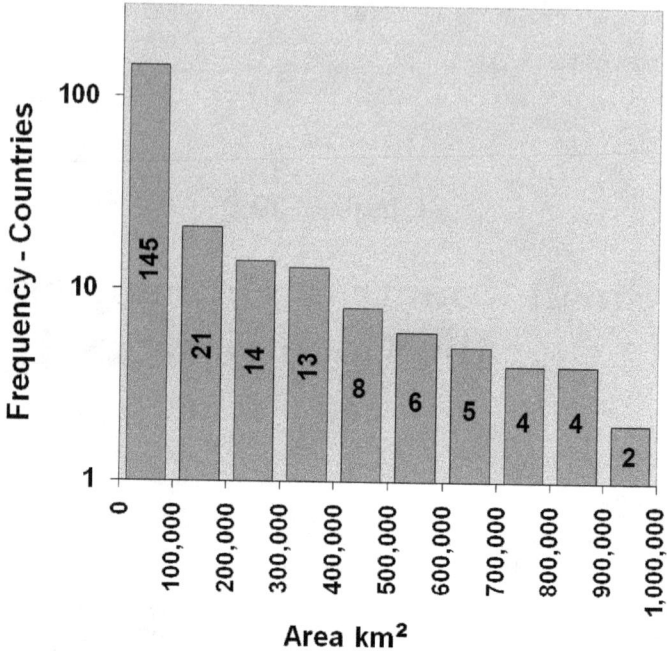

Figure 1.27:   Small Countries Decisively Outnumber Big Counties in the World

Russia is followed by Antarctica, Canada, China, the United States, Brazil, Australia, India, and Argentina, in that order.

Figure 1.27 depicts the histogram of the areas of 222 countries, from the smallest area size of 0.49 of the Vatican City imbedded in Italy, to area size of just below one million square kilometers, which is Egypt with 995,450 square kilometers. These 222 chosen values, out of the total of 253 countries in the data set, is missing only 31 of the biggest countries in the world with over one million square kilometers, whose presence over the x-axis is diluted and thinly spread out, getting rarer as we move along the x-axis in the positive direction of bigger areas. The histogram is constructed in even steps of 100,000 square kilometers for each successive bin. The vertical axis uses the logarithmic scale for better visualization, although this masks the sharp fall in the histogram, as it visually suppresses the enormous preference for the small which is indeed much more acute and dramatic as compared to what is seen in the histogram of Figure 1.27.

Some eccentric historians claim that small countries typically fear big countries, and for good reason, so they claim, but other more astute historians passionately refuse this claim, and bring up numerous counter examples from ancient as well as more modern history accounts.

Chapter 21

Small Prime Numbers Slightly Outnumber Big Prime Numbers

A prime number is an integer which cannot be expressed as a product of two smaller integers. Hence the only way to express a prime number P through some kind of a multiplication of two integers is P times 1. Integer 21 for example is not a prime number because it can be written as $(7) \times (3)$. Even though integer 1 really satisfies the definition of a prime, yet we arbitrarily and conveniently choose not to regard 1 as a prime number because this approach makes the setup and the writing of lots of theorems about the prime numbers much easier. For this reason, the smallest prime number is integer 2.

The set of all the prime numbers between 2 and 1000 contains 168 integers and it is outlined as follows:

2	3	5	7	11	13	17	19	23	29	31	37
41	43	47	53	59	61	67	71	73	79	83	89
97	101	103	107	109	113	127	131	137	139	149	151
157	163	167	173	179	181	191	193	197	199	211	223
227	229	233	239	241	251	257	263	269	271	277	281
283	293	307	311	313	317	331	337	347	349	353	359
367	373	379	383	389	397	401	409	419	421	431	433
439	443	449	457	461	463	467	479	487	491	499	503
509	521	523	541	547	557	563	569	571	577	587	593
599	601	607	613	617	619	631	641	643	647	653	659
661	673	677	683	691	701	709	719	727	733	739	743
751	757	761	769	773	787	797	809	811	821	823	827
829	839	853	857	859	863	877	881	883	887	907	911
919	929	937	941	947	953	967	971	977	983	991	997

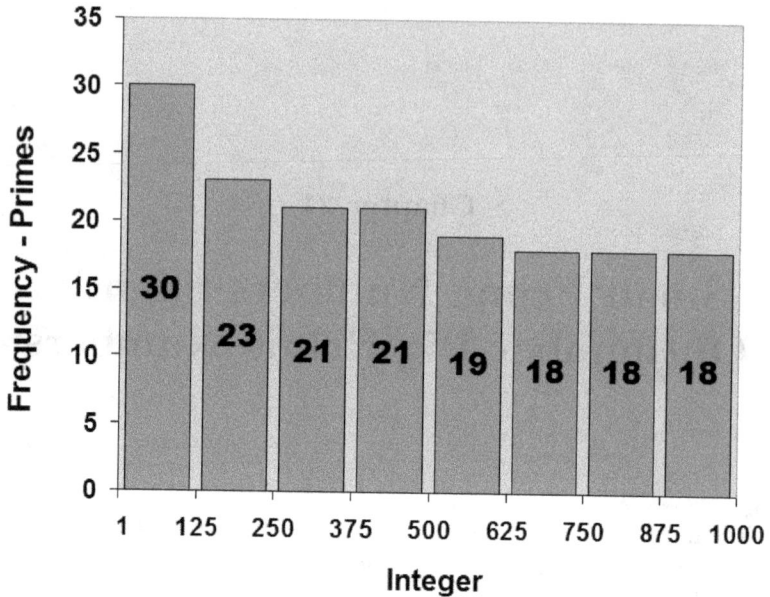

Figure 1.28:   Slightly More Small Primes than Big Primes — Integers 2 to 1000

The histogram in Figure 1.28 depicts the mild fall in the concentration of these 168 prime numbers. In the example of prime numbers, the small is not much more numerous than the big, in sharp contrast to all the earlier examples where dramatic difference in frequency were observed in favor of the small. Moreover, as consideration moves to higher primes, the small and the big obtain nearly the same frequencies, endowing the small only a minor advantage over the big; and even this mild advantage practically disappears at infinity.

Let us dramatically expand the scope of the above histogram from a termination at 1000 which yields only 168 primes, to a termination at 1,000,000 which yields as many as 78,498 primes. The histogram in Figure 1.29 depicts the spread of these 78,498 prime numbers from 2 to 1,000,000.

Visually, this last histogram of Figure 1.29 appears to convey milder differentiation between the small and the big — as compared with the histogram of Figure 1.28 where differences seem a bit more pronounced. Let us carefully quantify this vague visual observation

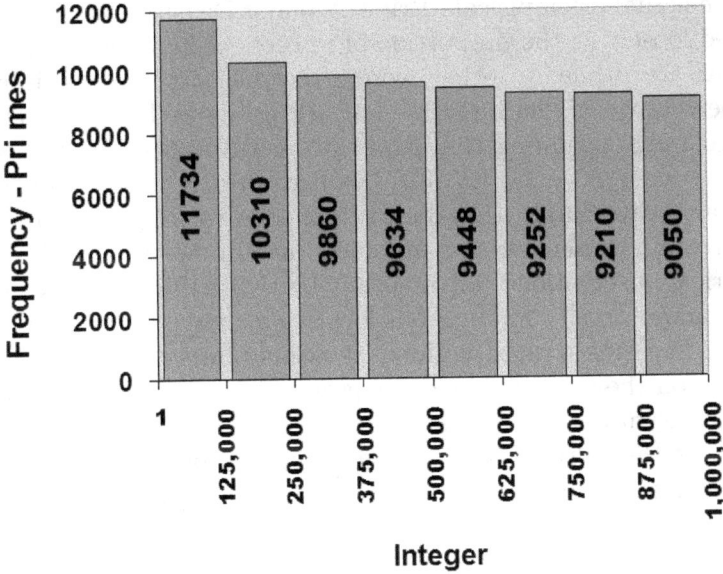

Figure 1.29: Lesser Advantage for the Small in Primes — 2 to 1,000,000

via the ratio of the height of the first bin to the height of the last bin; ignoring all the bins in the middle.

For the set of primes from 2 to 1000, the ratio Small/Big = 30/18 = **1.67**.
For the set of primes from 2 to 1 million, the ratio Small/Big = 11734/9050 = **1.30**.

Clearly, the small is losing its advantage as higher ranges of the integers are considered. In the limit — as much higher ranges of integers are considered — the above bin-ratio tends to 1.0 from above.

The case of prime numbers stands apart in the context of the small is beautiful phenomenon.

The forces at play here which are responsible for the particular spread of primes among the integers are completely distinct from the causes and explanations regarding most of the other manifestations of the small is beautiful phenomenon in real-life cases and data sets. Let us briefly examine the theoretical basis for the mildly skewed distribution of prime numbers as seen in Figures 1.28 and 1.29.

The actual or empirical "Prime Number Density" at integer N is defined loosely as the proportion of integers centered around integer N that are prime. In other words, the "density" refers simply to the percentage of the integers that are prime within an arbitrarily defined range spanning the immediate neighborhood of N, namely the interval $((N - I), (N + I))$, where integer I is chosen in an arbitrary way. Surely this density depends to some extent on the actual value chosen for I, namely on the size of the neighborhood. Empirical observations of prime numbers distribution show chaotic and random density fluctuations locally on short intervals, but quite orderly and stable density globally on long intervals, falling almost steadily on the right as higher rangers of integers are considered.

The asymptotic law of distribution of prime numbers, also known as the "Prime Number Theorem", theoretically predicts an approximate value for the empirical density as follows:

$$\text{Prime Number Density at } N \approx \frac{1}{\ln(N)}$$

For example, theoretical density predictions for N values of 100, 1000, 10000, 100000, 1000000, are $1/\ln(100)$, $1/\ln(1000)$, $1/\ln(10000)$, $1/\ln(100000)$, $1/\ln(1000000)$, respectively, namely the proportions 0.217, 0.145, 0.109, 0.087, 0.072, or 21.7%, 14.5%, 10.9%, 8.7%, 7.2% as percentages. Empirical examinations of actual prime number density values very closely match the above theoretical expectations; confirming the expression of the Prime Number Theorem.

This result clearly explains what was observed in Figures 1.28 and 1.29. Firstly, the theoretical local percentage of primes is inversely proportional to the logarithm of N, hence it decreases consistently and monotonically, as seen in the two histograms. Secondly, $1/\ln(N)$ flattens at infinity, and at least it tends to become flatter as higher values of integers are considered, and this fact explains why Figure 1.29 appears flatter than Figure 1.28.

Chapter 22

Four-Dice Multiplication Game
at the Casino in Favor of the Small

A dishonest casino is interested in providing a new and exciting game to attract more customers, with the conscious intention of confusing and distracting the gamblers so as to cause them to lose money. The new game involves the simultaneous throwing of four normal dice, each with six sides. The resultant four numbers, namely the four faces of the four dice, are then all multiplied by each other, and this product constitutes the variable determining whether gamblers win or lose. Figure 1.30 illustrates these four dice in the casino about to be thrown in the game.

The following four-dice multiplication examples help in demonstrating the game:

$$\{1, 1, 1, 1\} \rightarrow (1) \times (1) \times (1) \times (1) \rightarrow 1 \qquad \text{(the minimum)}$$

$$\{1, 1, 4, 1\} \rightarrow (1) \times (1) \times (4) \times (1) \rightarrow 4$$

$$\{2, 1, 2, 1\} \rightarrow (2) \times (1) \times (2) \times (1) \rightarrow 4$$

$$\{1, 6, 5, 1\} \rightarrow (1) \times (6) \times (5) \times (1) \rightarrow 30$$

$$\{6, 3, 2, 1\} \rightarrow (6) \times (3) \times (2) \times (1) \rightarrow 36$$

$$\{3, 4, 5, 5\} \rightarrow (3) \times (4) \times (5) \times (5) \rightarrow 300$$

$$\{6, 6, 6, 6\} \rightarrow (6) \times (6) \times (6) \times (6) \rightarrow 1296 \qquad \text{(the maximum)}$$

It is decided that the casino wins only if the product is small, and that gamblers win whenever the product is either medium or big. The abstract and generic size term "small" is defined here as

Figure 1.30:   Multiplication Game of Four Dice

the relatively narrow range of 1 to 300. The abstract and generic size terms "medium" and "big" are defined here collectively as the relatively wider range of 301 to 1296. On the face of it, the rule for this game appears reasonable and attractive to gamblers, as it allocates wider range for the gamblers, and narrower range for the casino. Yet, the high expectation on the part of naïve gamblers to earn money in this rigged game would surely give way to deep disillusionment and a substantial loss of money. Almost all bets would be won by the casino, as the vast majority of these four-dice combinations yield a product less than 300. Figure 1.31 depicts in details 50 such games for analysis — obtained via computer simulations.

Out of 50 such games, 41 came out smaller than 300, and only 9 came out bigger than 300. It should be noted that there is no generic preference for the small in any way for these four dice themselves. Indeed, these 200 individual dice throws shown in Figure 1.31 came out approximately with equal proportions for all 6 faces. Face 1 occurred 34 times; face 2 occurred 35 times; face 3 occurred 34 times; face 4 occurred 34 times; face 5 occurred 31 times; and face 6 occurred 32 times. All this is quite reasonable; since the theoretical expectation for each face in 200 throws assuming an unbiased die is $200/6 \approx 33$ times.

Figure 1.32 depicts the histogram of the products of those 50 games. The width of the first four bins is 100, but the widths of the last three bins are 200, 200, and 300; and not all bins are drawn exactly according to their true dimensions in the histogram. Certainly, the small decisively outnumbers the big in this game. It would be very difficult to argue that the small just happened to be more numerous here because of some supposed uniqueness in the numbers 6 and 4, or that somehow the combination of the value

Die 1	Die 2	Die 3	Die 4	Product
1	4	2	4	32
1	2	3	6	36
1	1	3	1	3
1	3	2	4	24
1	1	2	2	4
2	3	6	3	108
2	6	4	5	240
2	3	4	5	120
2	6	5	2	120
2	6	4	2	96
2	6	4	2	96
1	2	1	1	2
2	1	5	3	30
2	3	3	2	36
2	3	1	3	18
2	6	1	2	24
1	6	6	4	144
2	4	3	1	24
4	5	5	6	600
3	4	5	4	240
3	1	6	6	108
3	6	5	4	360
3	2	2	1	12
3	1	3	3	27
3	1	5	1	15
3	6	6	6	648
5	6	6	6	1080
3	2	5	5	150
3	3	5	1	45
3	6	5	6	540
4	4	5	1	80
4	5	5	1	100
4	3	4	5	240
4	5	3	1	60
4	5	1	4	80
4	4	2	1	32
4	6	3	1	72
5	6	3	4	360
5	4	4	1	80
5	1	1	2	10
5	1	4	6	120
5	2	3	2	60
5	5	2	1	50
5	4	6	3	360
6	4	4	2	192
6	4	3	3	216
6	5	3	6	540
6	5	4	5	600
6	2	1	2	24
6	4	2	2	96

Figure 1.31: Computer Simulations of 50 Four-Dice Games

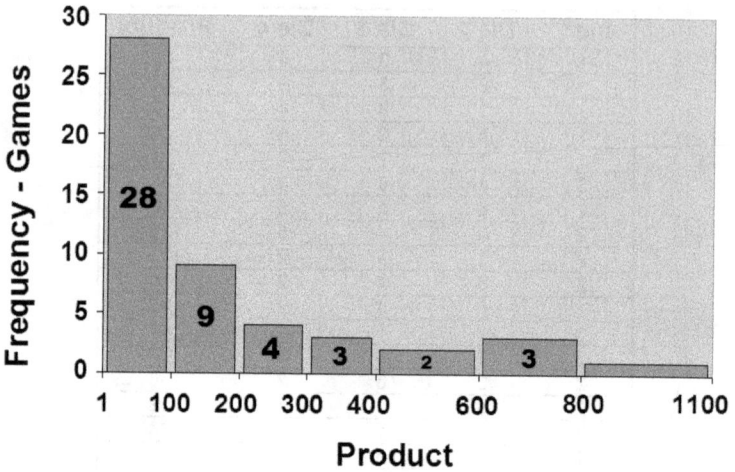

Figure 1.32:   Fifty Four-Dice Multiplication Games — Small Is Beautiful

of 6 (for a six-sided normal die) and the value of 4 (for these four dice) yields such uneven size results due to some rare numerical interaction. Surely there is nothing special about the numbers 6 and 4 or their combination, and that therefore the moral of the story in this four-dice multiplication game is that the relatively small is by far more numerous than the relatively big in all other types of multiplication processes.

Chapter 23

Three-Dice Selection Game at the Casino in Favor of the Small

Another game of complex dice dependencies was devised by the casino in the hope that even experienced and clever gamblers would get confused and lose the bet.

The first stage in the game is the throwing of a regular die of six sides (called "the primary die"). The second stage is the selection of the type of die to play next (called "the secondary die").

If 1 or 2 is obtained on the primary die then the secondary die is a four-sided one; if 3 or 4 is obtained on the primary die then the secondary die is an eight-sided one; if 5 or 6 is obtained on the primary die then the secondary die is a twelve-sided one. The third stage in the game is simply the throwing of the chosen secondary die (with four, eight, or twelve sides) to obtain the final value in the game. Figure 1.33 depicts the schematic dice arrangement of the game.

The following outline also illustrates the schematic dice arrangement:

[Primary Die] → [Secondary Die] → [Final Value]

The face value of the primary die determines the type [size] of die to be played as the secondary one, which is then thrown yielding the final value in the game.

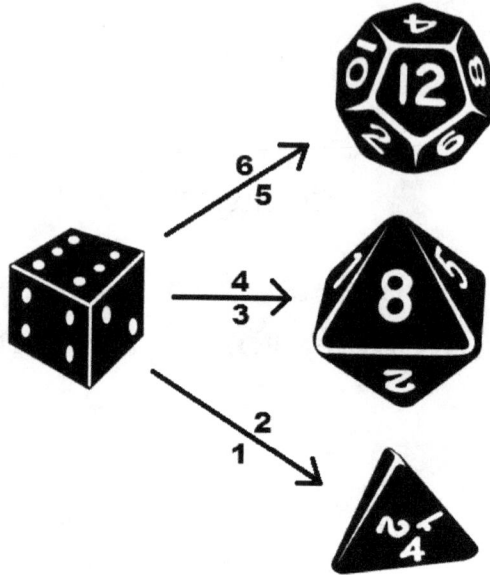

Figure 1.33:   The Throw of a Six-Sided Die Decides the Next Die to Play

The casino's rule for this game is deliberately set so as to appear highly generous to players.

Casino wins on $\{1, 2, 3, 4, 5\}$

Gamblers win on $\{6, 7, 8, 9, 10, 11, 12\}$

Since the generic idea of equal probabilities for all sides is strongly associated with games of dice, the mistaken assumption of the gamblers here is that all 12 possible occurrences come with equal probabilities; therefore the apparent generous offer by the casino to allow gamblers to win on seven occurrences as compared with the casino's ability to win only on five occurrences, entices many naïve gamblers to play the game. This conceptual argument is based on two equalities.

The first equality is that of the primary six-sided die which yields the same probability for all six sides, and by implication the same probability of $\{1, 2\}$, $\{3, 4\}$, $\{5, 6\}$ occurrences, leading to equal probability in the selection of the secondary die. The second equality is that of each of the four-sided, eight-sided, and twelve-sided dice,

6	→	Die 12	→	1	5	→	Die 12	→	7	6 → Die 12 → 2
1	→	Die 4	→	4	4	→	Die 8	→	5	5 → Die 12 → 12
1	→	Die 4	→	1	1	→	Die 4	→	1	6 → Die 12 → 9
1	→	Die 4	→	4	5	→	Die 12	→	10	4 → Die 8 → 1
2	→	Die 4	→	1	4	→	Die 8	→	7	5 → Die 12 → 6
3	→	Die 8	→	8	4	→	Die 8	→	5	4 → Die 8 → 6
3	→	Die 8	→	5	4	→	Die 8	→	4	2 → Die 4 → 3
2	→	Die 4	→	1	3	→	Die 8	→	4	4 → Die 8 → 4
6	→	Die 12	→	11	5	→	Die 12	→	2	5 → Die 12 → 10
1	→	Die 4	→	2	2	→	Die 4	→	1	2 → Die 4 → 2
6	→	Die 12	→	12	6	→	Die 12	→	1	2 → Die 4 → 3
1	→	Die 4	→	2	4	→	Die 8	→	2	3 → Die 8 → 5
6	→	Die 12	→	8	2	→	Die 4	→	3	1 → Die 4 → 3
6	→	Die 12	→	3	3	→	Die 8	→	7	1 → Die 4 → 3
4	→	Die 8	→	6	6	→	Die 12	→	11	5 → Die 12 → 5
1	→	Die 4	→	2	4	→	Die 8	→	3	5 → Die 12 → 1
5	→	Die 12	→	3	5	→	Die 12	→	1	2 → Die 4 → 4
4	→	Die 8	→	6	5	→	Die 12	→	3	3 → Die 8 → 6
5	→	Die 12	→	12	2	→	Die 4	→	2	1 → Die 4 → 2
5	→	Die 12	→	6	4	→	Die 8	→	2	1 → Die 4 → 3
6	→	Die 12	→	5	2	→	Die 4	→	2	6 → Die 12 → 1
6	→	Die 12	→	8	6	→	Die 12	→	2	6 → Die 12 → 1
1	→	Die 4	→	1	6	→	Die 12	→	9	5 → Die 12 → 6
3	→	Die 8	→	1	2	→	Die 4	→	3	2 → Die 4 → 3
1	→	Die 4	→	1	1	→	Die 4	→	3	2 → Die 4 → 1
5	→	Die 12	→	1	6	→	Die 12	→	8	3 → Die 8 → 1
4	→	Die 8	→	7	6	→	Die 12	→	9	5 → Die 12 → 9
1	→	Die 4	→	3	2	→	Die 4	→	3	2 → Die 4 → 3
1	→	Die 4	→	1	4	→	Die 8	→	1	1 → Die 4 → 2
4	→	Die 8	→	4	3	→	Die 8	→	4	1 → Die 4 → 3
2	→	Die 4	→	1	1	→	Die 4	→	4	5 → Die 12 → 5
1	→	Die 4	→	2	4	→	Die 8	→	3	3 → Die 8 → 1
5	→	Die 12	→	2	3	→	Die 8	→	2	
5	→	Die 12	→	10	3	→	Die 8	→	8	

Figure 1.34: Computer Simulations of 100 Three-Dice Selection Games

all being fair and unbiased, with equal probabilities for all possible sides.

As it turned out, most of the bets are actually won by the casino, as the majority of the final values in the game come out smaller than 6. Figure 1.34 depicts in detail 100 such dice selection games for analysis — obtained via computer simulations.

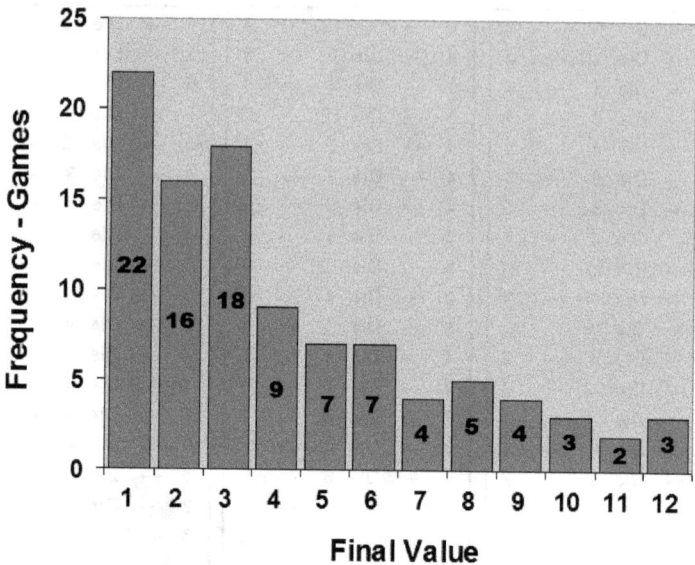

Figure 1.35:   Hundred Three-Dice Selection Games — Small Is Beautiful

Figure 1.35 depicts the histogram of the final values of those 100 simulated games. Clearly the small is more numerous than the big here albeit in a somewhat irregular and zigzag manner, as the overall trend is decisively in favor of the small. One could conjecture with total confidence that had many more games been simulated, say 10,000 games, then the histogram would be nearly perfectly smooth, falling monotonically and steadily to the right.

It should be noted that these 100 throws of the six-sided die in Figure 1.34 show no preference for the small at all. Indeed, these 100 throws came out with nearly equal proportions for all six faces. Face 1 occurred 20 times; face 2 occurred 16 times; face 3 occurred 12 times; face 4 occurred 16 times; face 5 occurred 19 times; and face 6 occurred 17 times. The theoretical expectation for each face in 100 throws assuming an unbiased die is $100/6 \approx 17$ times. In addition, the occurrences of each of the four-sided, eight-sided, and twelve-sided dice also seem even and balanced, and without any preference for the small. Yet, in spite of such evenness and equality within each die and at each stage in the process, the small decisively outnumbers the big in the overall game.

Chapter 24

Proper Criteria for Sizes Balancing the Chances of Big and Small

We cannot allow the criteria for sizes to be loosely defined, lest all measurements of frequencies and size dominance become arbitrary and meaningless. Statisticians and scientists need to heed not only the warning on the need to focus on relative quantities as opposed to absolute quantities, namely that size definitions should be made only relative to the data set under consideration, but they also need to heed another warning about fairness and balance in defining sizes within the context of the given data set, namely not to give larger range for one size at the expense of another size. Definitions and criteria for sizes closely relate to histogram construction in terms of the setup of bin-widths and bin-ranges, and the total numbers of bins in the histogram. If we are to interpret histograms correctly as to which sizes are dominant and which sizes are not, then these histograms must be constructed properly.

In order to illustrate the ramification of size definitions to conclusions regarding relative sizes, an imaginary revenue data on one particular day for a hypothetical small company shall be discussed.

Figure 1.36 depicts 90 fictitious sales amounts for this hypothetical company, sorted from low to high, and considering only whole dollar amounts without the insignificant cents.

Figure 1.37 depicts a crude histogram with only three thick bins for this fictitious revenue data. The company is of a very modest size and its clientele is small and poor, hence only 90 consumers purchased

13	309	662	873	1396	1839
25	322	670	897	1406	1867
37	338	686	912	1452	1910
41	359	695	936	1481	2007
49	403	717	955	1525	2026
52	422	724	964	1537	2053
55	438	728	972	1578	2075
61	459	740	1016	1592	2332
75	470	771	1057	1613	2358
82	473	783	1081	1657	2419
115	486	798	1108	1711	2465
123	492	809	1235	1737	2506
179	502	813	1334	1744	2652
204	638	862	1365	1751	2713
281	655	866	1387	1807	2975

Figure 1.36:   Ninety Fictitious Sales Amounts of a Hypothetical Company

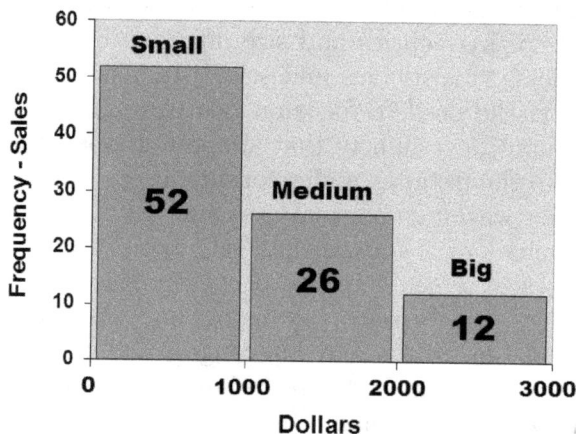

Figure 1.37:   Small Outnumbers Big in Fictitious Revenue Data

products on that day, and not a single sale above $3000 took place. Reasonable and fair histogram construction requires that all the bins are of equal width on the horizontal axis. Accordingly, the three bins of the histogram in Figure 1.37 are all with equal $1000 width.

A reasonable classification here would assign size names as follows:

SMALL	Sales from	$1	up to	$1000, namely [**1, 1000**)
MEDIUM	Sales from	$1000	up to	$2000, namely [**1000, 2000**)
BIG	Sales from	$2000	up to	$3000, namely [**2000, 3000**)

Therefore according to this classification, there are 52 small sales; 26 medium sales; and only 12 big sales. It can be said that for this fictitious revenue data, the small is more numerous than the medium, and the medium is more numerous than the big.

The standard type of histograms as seen in Figure 1.37 fairly allocates an identical width for each bin, enabling small, medium, and big to compete fairly for supremacy. Establishing uniformity in the width of the bins is usually an essential requirement for judging correctly from the shape of the histogram the relative occurrences of sizes; namely which sizes are more frequent in comparison having tall bins, and which sizes are less frequent in comparison having short bins; although in some particular cases of histogram construction, a mixture of some wide (thick) bins and some narrow (thin) bins is a useful procedure in order to obtain better visualization, but such an arrangement could potentially distort and confuse size proportion analysis.

The eccentric manager at the sales department of the hypothetical company treats the salespersons in a dictatorial manner, often accusing them of being too soft in negotiating prices. His long and thick moustache, his perpetually worn black beret as his office attire, and his alternating hoarse and husky voices, are all enough to intimidate even the boldest employee. He has formed his own highly opinionated notion of what should be considered "small", "medium", and "big", and would not allow anybody in the office to argue against his arbitrary benchmarks. Luckily for most of the salespersons at his office, his benchmarks schedule actually happened to be highly generous to them in the overall scheme of things, and they are happy about it. He harshly reprimands any salesperson returning with a sale transaction of less than $500, mocking it as "small" and as a failure, urging the salesperson to achieve better results and to sell more. On the other hand, he feels highly enthusiastic about any sale over the psychologically significant threshold of $1000, and considers such transactions as "big", successful, definitely worthy of praise and higher commissions. His size criterion can be summarized as follows:

SMALL	Sales from	$1	up to	$500, namely	[**1, 500**)
MEDIUM	Sales from	$500	up to	$1000, namely	[**500, 1000**)
BIG	Sales from	$1000	up to	$3000, namely	[**1000, 3000**)

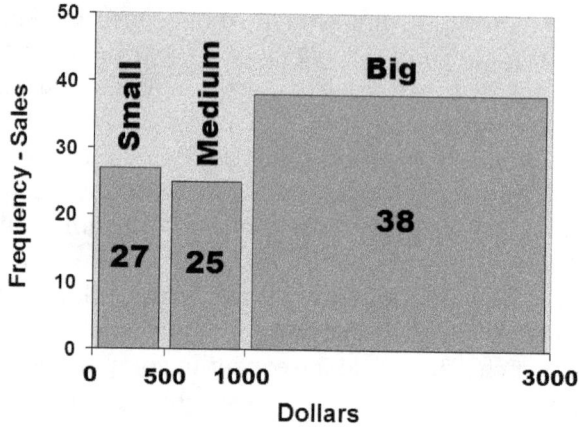

Figure 1.38:   Distortion of True Size Proportions Due to Unequal Bin Widths

The eccentric manager then sketched his own self-styled histogram regarding the fictitious revenue data of Figure 1.36 by using his own benchmarks for sizes, as shown in Figure 1.38.

Hence, according to his eccentric notions of sizes, the big is the most numerous. The histograms of Figures 1.37 and 1.38 compete in how to best represent the same data set of Figure 1.36, and these two histogram arrangements arrive at profoundly different conclusions about size proportions. Surely, such artificial and arbitrary definitions of sizes and bin widths of the eccentric manager distort true size proportions, and should not be accepted as a good presentation of the company's sales data.

Given a standard bin-width applied equally to all bins, then a consistently and monotonically falling histogram (onto the right side) implies that smaller quantities are more numerous and that bigger quantities are relatively rare. A flat and horizontal histogram implies that all the sizes are equally significant. A consistently and monotonically rising histogram (on the right side) implies that the big is more numerous than the small throughout the entire range of the data.

In many of the examples and cases of this section of the book, the bins on the horizontal x-axis were of unequal sizes, in violation of what is recommended in this chapter. Yet, this was done in order to improve visualization of the data by making the big a bit visible and saving it from total obscurity, knowing in advance that the big would earn only very few extra data points — as the small was overwhelming the big a great deal under the assumption of equality of bin widths.

In other words, several bins of extremely low frequency counts belonging to the big on the right part of the histogram were allowed to be merged, knowing a priori that such a consolidation by the big would not overturn the dominance of the small in any way whatsoever.

The use of the logarithmic scale for the horizontal axis should also be carefully considered, as it is proper and even essential in some cases, but wrong and misleading in others. In this example of revenue data of the hypothetical company, the use of a logarithmic horizontal scale would be utterly misguided and would totally distort true size configuration. The log of the minimum 13 is 1.1139, and the log of the maximum 2975 is 3.4734, hence the partitioning of the entire log range into three equal sections yields $(3.4734 - 1.1139)/3 = 0.7865$ as the width for each bin, implying that the three log sub-intervals for small, medium, and big should be $(1.1139, 1.9004)$, $(1.9004, 2.6870)$, $(2.6870, 3.4735)$ respectively. Examining the number of log-data points falling within these three sub-intervals gives the count of 9 for small, 17 for medium, and 64 for big; and surely this is an utterly distorted and inverted vista of size configuration here. Figure 1.39 depicts the inverted vista of size configuration

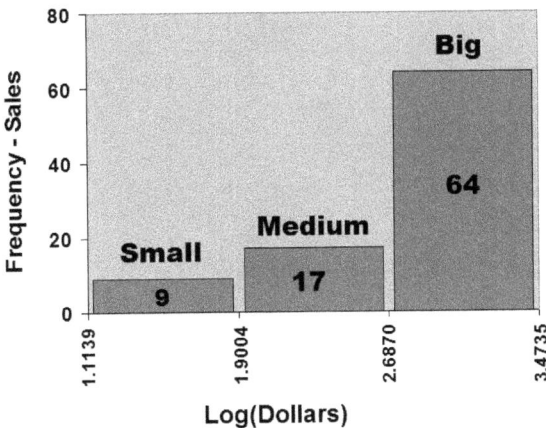

Figure 1.39: Distortion of True Size Proportions Due to Horizontal Logarithmic Scale

due to the misguided use of the logarithmic scale for the sales values. Some additional insight can be gained here by considering how these three log sub-intervals look like when translated back into their numerical equivalents, which are the sub-intervals $(13, 80)$, $(80, 486)$, $(486, 2975)$. These three numerical sub-intervals were carved out of the entire $(13, 2975)$ range in an extremely unbalanced way, endowing them the very distinct widths of 67, 407, 2488, respectively. This fact together with a casual reading of the table in Figure 1.36 explains the unexpected rise of the big and the sudden fall in the small.

Chapter 25

Exceptions and Counter Examples to the Phenomenon

Not all things in the world are quantitatively configured in favor of the small. There are counter examples and exceptions. But *most* things in the world and the *majority* of empirical data sets are with a decisive preference for the small.

The abstract theoretical "Normal Distribution" in mathematical statistics serves as a significant template and a model of one such exception. The Normal Distribution is the antithesis of the small is beautiful principle. The histogram of the Normal Distribution is symmetrical, and it appears as a bell shape, starting low on the left for small values, rising sharply to the maximum height in the middle for medium values, then falling precipitously on the right for big values.

Figure 1.40 depicts the histogram of 50,000 Monte Carlo computer simulations of the abstract Normal Distribution with a mean value of 9 and standard deviation value of 2. Here neither the small nor the big are beautiful, but rather the medium is such, as the histogram is symmetrical and most of the data points fall around the center.

The Three Sigma Rule in empirical statistics states that for many reasonably symmetric distributions with only one mode, 99.7% of the data lies within three standard deviations (sd) of the mean. Therefore in the approximate, for this particular Normal, nearly all data points occur on the x-axis interval of $[\text{mean} - (3) \times (\text{sd}), \ \text{mean} + (3) \times (\text{sd})] = [9 - (3) \times (2), \ 9 + (3) \times (2)] = [3, 15]$, and this range for the Normal here can be visually confirmed in Figure 1.40 in the approximate.

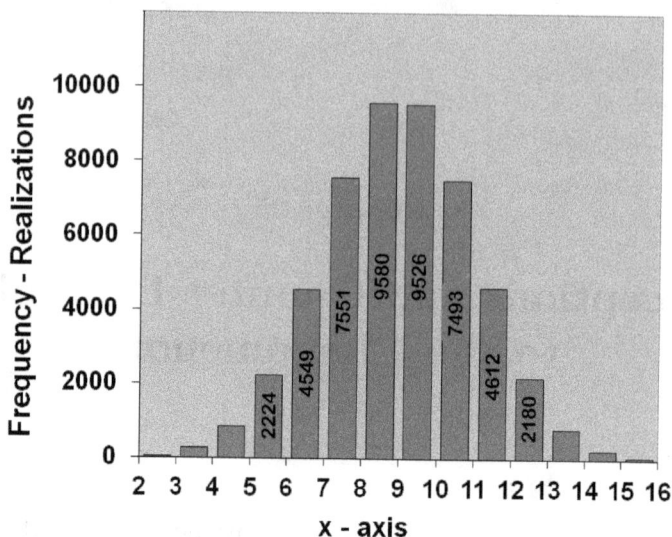

Figure 1.40: 50,000 Simulations of the Normal(9, 2) — Medium Is Beautiful

The height of people is one such rare case out of several other counterexamples and exceptions to the nearly universal small is beautiful phenomenon. Typical spread of height values of a large group of people closely resembles the Normal Distribution curve, favoring the medium over either the small or the big.

The weight of people is another notable exception to the small is beautiful principle, and which also resembles the Normal a bit, although the fit to the Normal for people's weight is not as good as the fit for people's height. The majority of people are neither too overweight nor extraordinarily thin; rather most people are of medium weight relative to their gender, age, and race.

The English Premier League (EPL) is the highest level of the English football league system. It is contested by 20 clubs or teams, and it operates on a system of promotion and relegation within the EPL. Seasons usually run from August to May. The University of Florida at Gainesville maintains a large database on a variety of real-life issues, including a detailed list of 526 players of the EPL, and it is available on their website under the title "EPL 2014-2015 Player Heights and Weights" via the link <https://users.stat.ufl.edu/~win ner/datasets.html>. The website publishes detailed information for each player regarding Team, Player Name, Player Number, Player

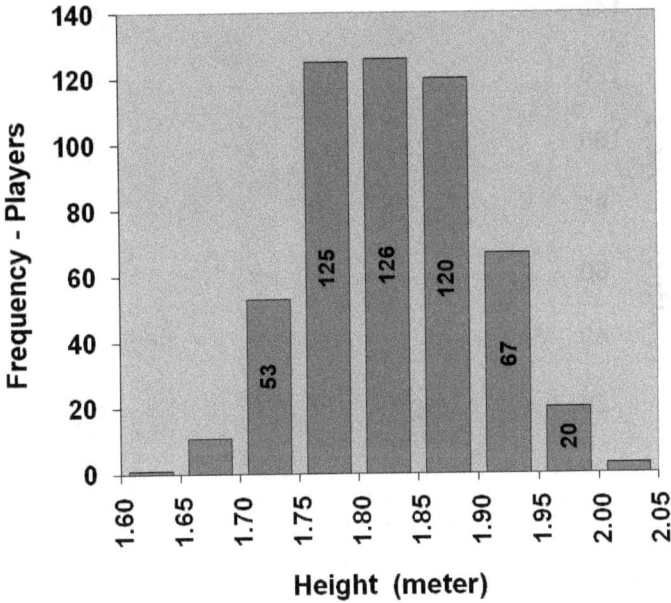

Figure 1.41: English Premier League, Player Height — Medium Is Beautiful

Position, Age, Height in inches, and Weight in pounds. The last two variables are converted into meter and kilogram units respectively, using the conversion rules that 1 inch equals 0.0254 meters, and that 2.20462 pounds equal 1 kilogram.

Figure 1.41 depicts the histogram of the heights of these 526 EPL players using the unit of the meter.

Figure 1.42 depicts the histogram of the weights of these 526 EPL players using the unit of the kilogram.

The height histogram appears to be a bit closer to the Normal Distribution and more symmetrical, as compared with the weight histogram which appears with less fit to the overall shape of the Normal, as it is a bit asymmetrical.

Other notable exceptions which are also modeled on the Normal Distribution are:

Serum total cholesterol, HDL-cholesterol, and LDL-cholesterol blood concentrations.
Probability density function of a ground state in a quantum harmonic oscillator.

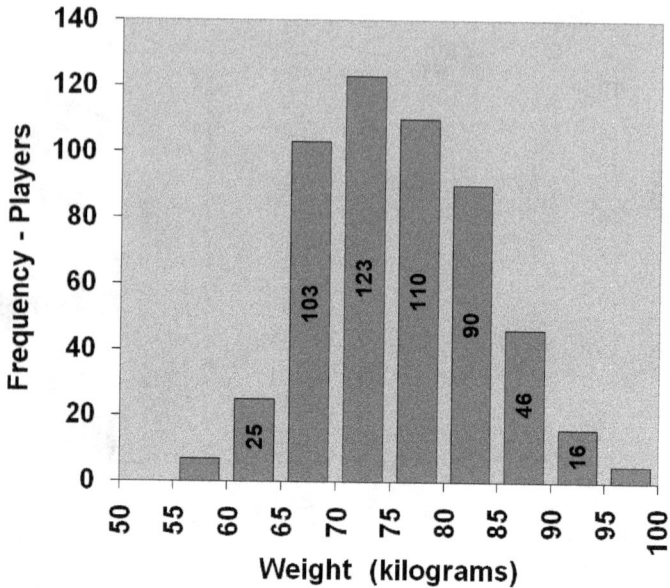

Figure 1.42:   English Premier League, Player Weight — Medium Is Beautiful

IQ test results, and some other standardized tests results.
Diastolic blood pressure. Heart rates.
Measurement errors in physical experiments.
Other certain (but very few) quantities in physics.
The position of a particle that experiences diffusion.

In all of these counter-examples above, the medium is the most frequent size, while the small and the big are rare in comparison. Yet, in spite of its great fame, and beyond these examples and several others, the Normal Distribution is not prevalent in real-life measurements and data sets.

The flat and horizontal "Uniform Distribution" is another antithesis of the small is beautiful principle, as it is quantitatively configured in such a way that all sizes and quantities are approximately equally likely. The Uniform Distribution is very rarely found in nature, and only very few real-life processes and cases have this unique and simplistic form of distribution.

The "birthday" date of a large group of people is one such rare example of the Uniform Distribution. In normal and peace times, and where weather and climate are not big factors, weddings and mating habits are not concentrated in any particular month or time. All dates within the 365 possible days within a year are equally likely for births, with neither preference nor bias against any month or week, or any specific period in the year. A person is just as likely to be born for example in early March, late November, or mid-February. Sizes in this context are related to "early" in the year versus "late" in the year, thus January which is very early in the year is thought of as very small; March which is a bit early in the year is considered as a bit small; while June or July are considered as medium; and December is considered as the biggest since it is the latest month in the year.

One real-life example is given with the detailed database on all Nobel Prize laureates during the entire history of the Nobel Foundation in Sweden since Prizes were first awarded in 1901 to

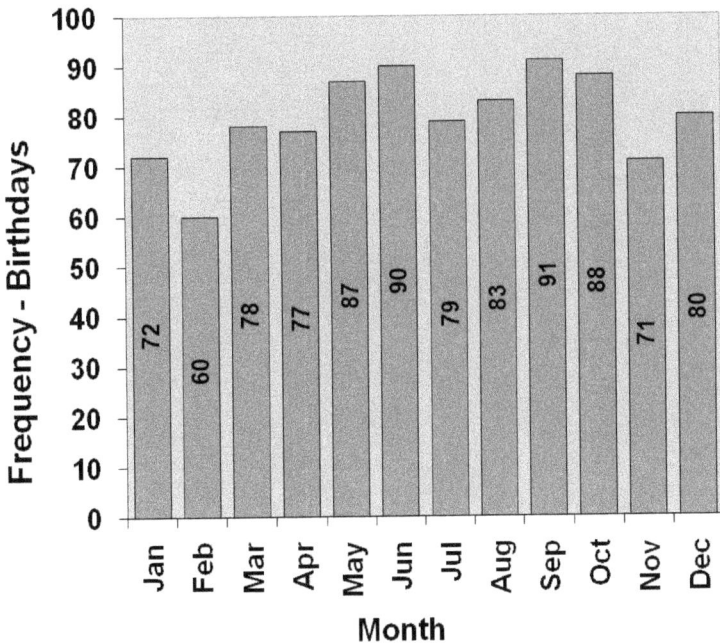

Figure 1.43: Birthdays of Nobel Laureates — None of the Sizes Are Beautiful

the present. The database includes the birthday information of almost all laureates. The link to the website is found at:

https://public.opendatasoft.com/explore/dataset/nobel-prize-laureates/table/?disjunctive.category

The website contains 956 listed laureates with information on their birthday. Figure 1.43 depicts the histogram of the birthdays of the 956 laureates by month. In the aggregate, from January to December, the histogram neither rises nor falls, and it does not have a definite bulge or dent around the center, therefore it appears to be shaped in the spirit of the Uniform, where none of the sizes are favored or discriminated against. The reason why the laureate's histogram does not appear flat enough is only due to the relatively small size of the data, being around only one thousand persons. For a big size random group of people with tens of thousands of persons, the birthday histogram would appear very nearly flat.

Chapter 26

Seven-Dice Addition Game at the Casino in Favor of the Medium

Another dishonest casino is interested in providing a new and exciting game to attract more customers, and with the conscious intention of confusing and distracting the gamblers so as to cause them to lose money. The new game involves the simultaneous throwing of seven normal dice, each with six sides. The resultant seven numbers, namely the seven faces of the seven dice, are then all added together, and this sum constitutes the variable determining whether gamblers win or lose. Figure 1.44 illustrates these seven dice in the casino about to be thrown in the game.

The following seven-dice addition possibilities help in demonstrating the game:

$\{1,1,1,1,1,1,1\} \rightarrow (1) + (1) + (1) + (1) + (1) + (1) + (1) \rightarrow \quad 7 \quad \text{(the minimum)}$
$\{3,2,1,2,1,3,1\} \rightarrow (3) + (2) + (1) + (2) + (1) + (3) + (1) \rightarrow \quad 13$
$\{4,1,2,3,1,4,3\} \rightarrow (4) + (1) + (2) + (3) + (1) + (4) + (3) \rightarrow \quad 18$
$\{3,5,5,2,1,3,4\} \rightarrow (3) + (5) + (5) + (2) + (1) + (3) + (4) \rightarrow \quad 23$
$\{5,4,4,5,6,5,5\} \rightarrow (5) + (4) + (4) + (5) + (6) + (5) + (5) \rightarrow \quad 34$
$\{6,6,6,6,6,6,6\} \rightarrow (6) + (6) + (6) + (6) + (6) + (6) + (6) \rightarrow \quad 42 \quad \text{(the maximum)}$

The casino determines the rules of the game, and it decrees that if the sum of all the seven dice falls in the narrow middle range between 20 and 30 the casino wins, while players win on the longer ranges from 7 to 19 as well as from 31 to 42. The abstract and generic size term "small" could be defined here as the range of 7 to 19, "medium" could be defined as the range of 20 to 30, and "big" could be defined as the range of 31 to 42.

Figure 1.44:   Addition Game of Seven Dice

Die 1	Die 2	Die 3	Die 4	Die 5	Die 6	Die 7	Sum
1	3	6	5	6	6	2	29
2	1	2	3	6	4	6	24
1	3	6	1	2	2	1	16
4	2	5	2	1	3	1	18
2	3	1	2	6	1	4	19
1	1	6	5	4	1	6	24
6	3	4	6	4	4	2	29
3	3	4	6	1	1	6	24
6	2	6	4	1	4	4	27
6	6	5	6	6	6	6	41
6	3	2	2	6	6	4	29
6	3	2	5	3	1	1	21
3	4	1	2	2	1	2	15
4	4	4	4	3	4	5	28
1	1	2	6	5	2	3	20

Figure 1.45:   Simulations of 15 Seven-Dice Random Games at the Casino

The Central Limit Theorem shall be discussed in a later chapter, but briefly stated the theorem predicts that additions lead to the Normal Distribution, hence, this sum of seven dice should resemble the Normal, and the dishonest casino knows this mathematical fact, so it utilizes the subtlety of the theorem and the naivety of the players to make sure it nearly always wins. The casino pretends to be fair and very generous by allocating greater overall range to the players.

In this game, the casino wins only if the sum is of medium size, and the gamblers win whenever the sum is of either small or big sizes. On the face of it, the rule for this game appears reasonable and attractive to gamblers, as it allocates for the gamblers more than double the range that it allocates for the casino. Yet, the high expectation on

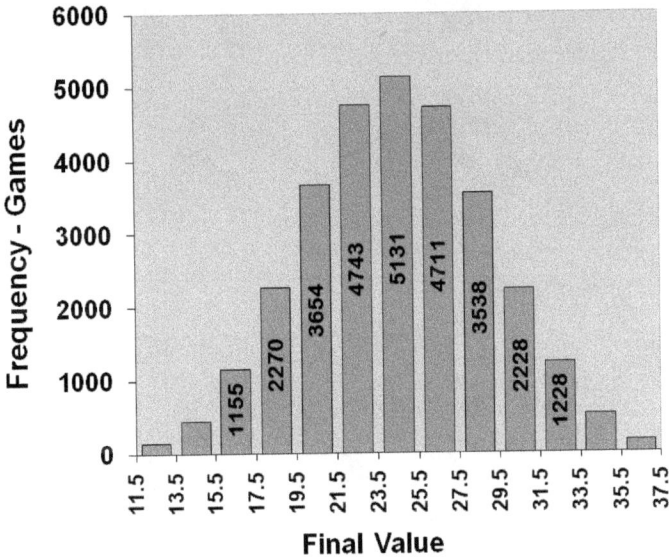

Figure 1.46: 30,000 Seven-Dice Addition Games — Medium Is Beautiful

the part of naïve gamblers to earn money in this rigged game would surely give way to deep disillusionment and substantial loss of money. Almost all bets would be won by the casino, as the vast majority of these seven-dice combinations yield a sum around the middle range which is termed "medium". Figure 1.45 depicts in detail 15 such games for demonstration, obtained via computer simulations. Figure 1.46 depicts the resultant histogram in 30,000 such computer simulations of the seven-dice game, greatly resembling the Normal Distribution as predicated by the Central Limit Theorem.

Chapter 27

Beauty Is in the Eye of the Beholder

Literally in English, the "beholder" is the person who holds the vision of the figure or the object in his or her mind, as visual information is derived from the eyes. The meaning of the phrase is that beauty cannot be judged objectively, only subjectively, for what one person finds beautiful or admirable may not appeal to others. In the numerical context of small and big sizes, one may investigate the extent by which the small is more numerous than the big; or may ask metaphorically: "By how much is the small more beautiful than the big?" The subsequent question that naturally arises in this context is whether or not there exists any consistent pattern for some numerical measurement designed to quantify the differences in size occurrences.

The profound challenge or dilemma prior to quantifying differences between sizes, is the necessity to have proper definition of what should constitute "small" and what should constitute "big". Should these two concepts be defined simply as the height of the first bin and the height of the last bin on the histogram, and the ratio Small/Big chosen as the numerical measure expressing the disparity in beauty? Certainly not! Histogram construction is arbitrary, subjective, and personal. Some data analysts construct histograms with many thin and refine bins, while others prefer only few thick and crude bins, and surely exact numerical comparison between the first bin (for small) and the last bin (for big) would depend on the style of histogram construction. Hence in the context of the small is beautiful phenomenon: "Beauty is in the eye of the histogram constructor".

Let us construct several histograms with varying bin width for the list of 2175 commonly used and naturally occurring chemical compounds of Chapter 3, and let us examine how relative frequencies of the first and the last bins are compared. The smallest molar mass in the data is 7.94894, and there are only 10 compounds with molar mass greater than 900, hence for all practical purposes all our histograms could start at 8 and terminate at 900 without any loss of generality.

Dividing the entire range of 8 to 900 into **2 equal parts** leads to $(8, 454)$ considered as "small" and having 1950 data points; and $(454, 900)$ considered as "big" and having 215 data points. This scheme yields the Small/Big ratio of $1950/215 = $ **9.07**.

Dividing the entire range of 8 to 900 into **3 equal parts** leads to $(8, 305.3)$ considered as "small" and having 1556 data points; $(305.3, 602.7)$ considered as "medium" and having 541 data points; and $(602.7, 900)$ considered as "big" and having 68 data points. This scheme yields the Small/Big ratio of $1556/68 = $ **22.88**.

Dividing the entire range of 8 to 900 into **4 equal parts** leads to $(8, 231.0)$ considered as "small" and having 1182 data points; $(231.0, 454.0)$ considered as "lower medium" and having 768 data points; $(454.0, 677.0)$ considered as "higher medium" and having 178 data points; and $(677.0, 900)$ considered as "big" and having 37 data points. This scheme yields the Small/Big ratio of $1182/37 = $ **31.95**.

Causes and Explanations of the Phenomenon

Chapter 28

Partitioning as a Cause of the Small Is Beautiful Phenomenon

The mathematical field of Integer Partitions investigates the ways an integral quantity can be expressed as the sum of (smaller) integers. It deals with very simple and straightforward questions such as: "In how many ways and exactly how can the quantity 7 be broken into integral parts?" The answer to this question is the exhaustive list of all possible integral partitions of 7 as shown in detail below:

7
$6 + 1$
$5 + 2$
$5 + 1 + 1$
$4 + 3$
$4 + 2 + 1$
$4 + 1 + 1 + 1$
$3 + 3 + 1$
$3 + 2 + 2$
$3 + 2 + 1 + 1$
$3 + 1 + 1 + 1 + 1$
$2 + 2 + 2 + 1$
$2 + 2 + 1 + 1 + 1$
$2 + 1 + 1 + 1 + 1 + 1$
$1 + 1 + 1 + 1 + 1 + 1 + 1$

If the focus is on the quantity 5 instead of the quantity 7, then the question posed is: "In how many ways and exactly how can the quantity 5 be broken into integral parts?" The answer to this question is the exhaustive list of all possible partitions of 5, and this is shown below together with some philosophical comments about sizes:

5   ← few big parts
4 + 1   ← few big parts
3 + 2
3 + 1 + 1
2 + 2 + 1
2 + 1 + 1 + 1   ← many small parts
1 + 1 + 1 + 1 + 1   ← many small parts

These two examples demonstrate a very profound, universal, and yet extremely simple principle regarding how a conserved quantity can be partitioned into parts, namely the observation that: "One big quantity is composed of numerous small quantities", or equivalently: "Numerous small quantities are needed to merge into one big quantity".

Partitioning a fixed quantity into parts can be done roughly-speaking in two extreme styles, either via a breakup into many small parts, or via a breakup into few big parts. For example, two extreme styles of integer partition of 7 are $\{1, 2, 1, 1, 1, 1\}$ with many small parts, and $\{3, 4\}$ with fewer but relatively bigger parts. A more moderate style perhaps would be to have a mixture of all kinds of sizes, consisting of many small ones, some medium ones, and few big ones. The above conceptual outline is one of the chief causes why so often real-life data sets are skewed quantitatively, having numerous small values, but only very few big values.

The principle merits additional visual presentations. Figure 2.1 demonstrates all possible integer partitions of 5, where the small clearly outnumbers the big in the entire scheme. Naturally, 3 could be designated as the middle-size quantity here. Consequently, 1 and 2 are designated as the small ones, while 4 and 5 are designated as the big ones. According to such classification of sizes, there are only 2 big quantities, but as many as 16 small quantities. In any case, one should not lose sight of the main aspect seen in Figure 2.1 which shows a mixture of all sorts of sizes, but with a strong bias towards the small, discriminating against the big. This is of course true in all Integer Partitions, and not only for integer 5. Figure 2.2 which organizes all the parts of Figure 2.1 nicely according to size,

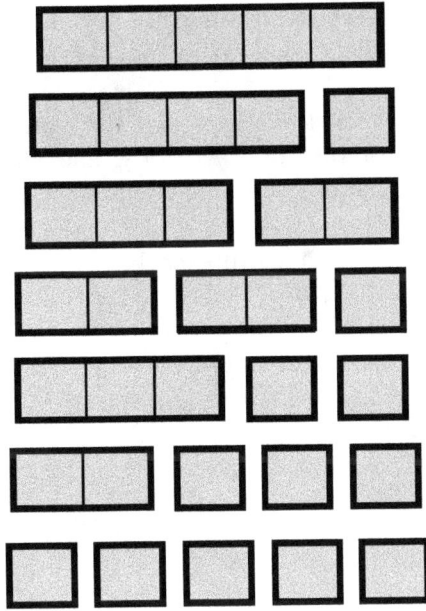

Figure 2.1:   The Small Outnumbers the Big in Integer Partitions of 5

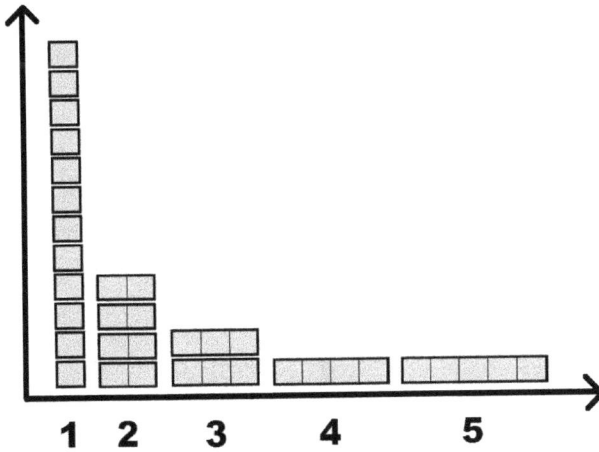

Figure 2.2:   The Organization of All Possible Integer Partitions of 5

clearly demonstrates the above principle, and if modified as a proper histogram then it can be said to be positively skewed with a tail falling on the right.

The abundance of small integers and the rarity of big integers is an intrinsic feature of Integer Partitions, and this is certainly not restricted to the examples of Integer Partitions of 7 and Integer Partitions of 5.

As an additional example, Integer Partitions of 13 is similarly analyzed. There are 101 distinct possible partitions for integer 13; as compared with only 15 possible partitions for integer 7, and only 7 possible partitions for integer 5. Only five partition examples out of the complete set of 101 possibilities are shown below as follows:

$$13 = 7 + 4 + 1 + 1$$
$$13 = 4 + 3 + 3 + 2 + 1$$
$$13 = 10 + 1 + 1 + 1$$
$$13 = 5 + 5 + 3$$
$$13 = 7 + 6$$

Carefully counting integer occurrences in all such possible partitions of 13 yields the following:

Integer	1	occurs	272	times.
Integer	2	occurs	112	times.
Integer	3	occurs	63	times.
Integer	4	occurs	38	times.
Integer	5	occurs	25	times.
Integer	6	occurs	16	times.
Integer	7	occurs	11	times.
Integer	8	occurs	7	times.
Integer	9	occurs	5	times.
Integer	10	occurs	3	times.
Integer	11	occurs	2	times.
Integer	12	occurs	1	time.
Integer	13	occurs	1	time.

Figure 2.3 depicts the histogram of *all* the integers from 1 to 13 within *all* 101 possible partitions. There are 556 integers in total residing within these 101 partitions of 13. The histogram is consistently and monotonically falling to the right, except at the very end for integer 12 with frequency 1 and for integer 13 with frequency 1. Since integer 12 and integer 13 occur exactly once in the entire scheme, namely as $13 = 13$ and as $13 = 12 + 1$, the histogram is

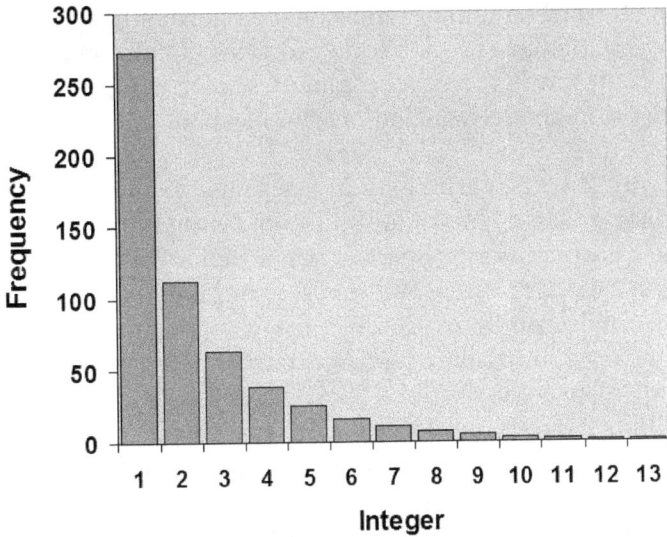

Figure 2.3:   All Possible Integer Partitions of 13 — Small Is Beautiful

actually flat and horizontal there for the tiny part on the right-most part of the histogram.

All histograms in Integer Partitions of whatsoever integer N are flat at the end of the tail on the right-most part for the largest two integers. This is so since $N = N$ and $N = (N - 1) + 1$.

Yet the histogram never retreats and it never rises, and thus in the case of Integer Partitions it can be said in general that the small is consistently more numerous than the big (with the irrelevant and insignificant exception of the largest two integers).

If we ask a friend to partition a given integral quantity into smaller integral parts only once, in any way he or she sees fit, then no grand conceptual principles can be applied. The friend could favor the big, or could favor the small. If possible, the friend might even partition the quantity into completely even pieces perhaps, where all the parts are of the same quantity, resulting in one size only. Clearly, no quantitative prediction can be made whatsoever for a single partition. Yet, if we ask the friend to randomly repeat partitioning many times over, or if he or she has the time and the patience to perform all possible partitions and to present them as one vast data set, then the configuration of the small is beautiful is inevitable! Such resultant blind bias toward the small is observed as long as he or she does not favor any particular sizes, treating all sizes equally and fairly. This generic bias towards the small is almost a universal principle in most other partition models, including those where fractional parts are allowed.

In Figure 2.4, the entire area in the shape of an oval representing the original value is partitioned in two distinct ways according to size. The first partition in the left panel divides the oval-shape area into big parts, and therefore the low number of only 5 parts is obtained. The second partition in the right panel divides the [same] oval-shape area into much smaller parts, and therefore the high number of 13 parts is obtained. Staring at Figure 2.4 reinforces the inevitability of the small is beautiful consequence in all partitions of a conserved quantity into parts having a variety of sizes. This territorial example broadens the quantitative application of Integer Partitions into fractional partitions as well.

Figure 2.5 provides another visual and intuitive demonstration that a given random partition containing the big, the small, and the medium; and where all sizes mix together randomly, many more small pieces should be found than big pieces. Figure 2.5 focuses on the area as the quantitative variable, but the lesson learnt from it is generic and applicable to any other types of quantities. Figure 2.5 depicts one possible random partition in the natural world where approximately 1/3 of the entire oval area consists of big parts (around the left side); approximately 1/3 of the entire oval area consists of small parts (around the center); approximately 1/3 of the entire oval area consists of medium parts (around the right side), namely endowing equal portions of overall quantity fairly to each size without any bias. Surely in nature there exists

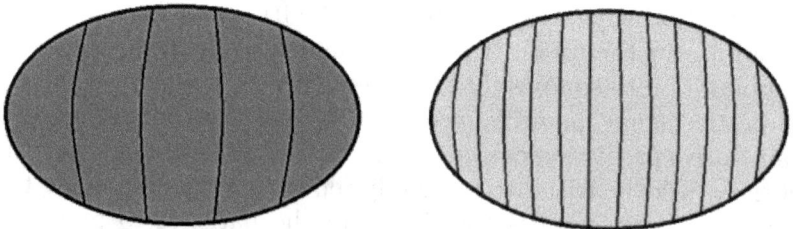

Figure 2.4: A Given Quantity Is Partitioned into Either Few Big Parts or Many Small Parts

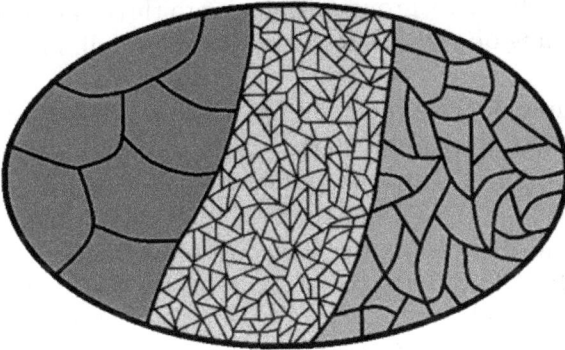

Figure 2.5:   An Equitable Mix of Small, Medium, and Big Yielding "Small Is Beautiful"

no such order and grouping by size around the left-center-right sections or along any other dimensions. In nature, the big, the small, and the medium are all mixed in chaotically. But the order for the sizes along the left-center-right sections shown in Figure 2.5 is made for pedagogical purposes, to reinforce visually for the reader the profound quantitative consequences affecting sizes in typical partition models. The example given via Figure 2.5 broadens the quantitative application of Integer Partitions into fractional partitions as well.

Hence, the data of Figure 2.5 is structured in such a way that adding all small values yields approximately the same sum as adding all big values (or medium). Yet, in spite of such equitable allocation of quantitative portions to the 3 sizes, the small outnumbers the big. Vaguely counting the number of enclosed areas for each size leads to the decisive conclusion that the small is by far more numerous than the median, and that the median is definitely more numerous than the big. Careful count of the enclosed areas in the oval shape above shows that there are **220** small parts in the middle, **35** medium parts on the right, and only **7** big parts on the left.

It should be emphasized that in fact, real-life data sets frequently deviate from the 33.3% fair allocation of overall quantitative portion for the three sizes. Also evidently, a great deal depends on the exact definition of what should constitute small, big, and medium. Since real-life data sets almost never occur nicely with exactly and merely 3 sizes, but rather mostly with numerous distinct values, it is necessary to arbitrarily group all values into three camps according to size — assuming one wishes to stick with the small, big, and medium categories.

For example, for the Integer Partitions of 5 discussed earlier, the large data set of all possible partitions is $\{5, 4, 1, 3, 2, 3, 1, 1, 2, 2, 1, 2, 1, 1, 1, 1, 1, 1, 1, 1\}$. The overall quantity is $(5+4+1+3+2+3+1+1+2+2+1+2+1+1+1+1+1+1+1+1) = 35$. Considering 1 and 2 as small and 4 and 5 as big, the portion of big is $(5 + 4)/(35) = (9)/(35) = \mathbf{25.7\%}$; the portion of small is $(1+2+1+1+2+2+1+2+1+1+1+1+1+1+1+1)/(35) = (20)/(35) = \mathbf{57.1\%}$; and the portion of medium is $(3+3)/(35) = (6)/(35) = \mathbf{17.1\%}$. Such a state of quantitative affairs is extremely in favor of the small, much more so than the 33.3% equal allocation for all three sizes. But surely, the definitions of small, big and medium here are arbitrary.

Considering small as $\{1\}$, big as $\{5\}$, and medium as $\{2, 3, 4\}$ also leads to the small is beautiful conclusion, although of slightly different intensity in beauty. Such re-definition of sizes also yields different allocation of portions from the 35 overall quantity. The portion of big is $(5)/(35) = \mathbf{14.3\%}$; the portion of small is $(1 + 1 + 1 + 1 + 1 + 1 + 1 + 1 + 1 + 1 + 1 + 1)/(35) = (12)/(35) = \mathbf{34.3\%}$; and the portion of medium is $(4 + 3 + 2 + 3 + 2 + 2 + 2)/(35) = (18)/(35) = \mathbf{51.4\%}$.

Figure 2.6 depicts another quantitative configuration that could also occur in the physical world. Here the small constitutes approximately 80% of the entire area (i.e. entire quantity), while the

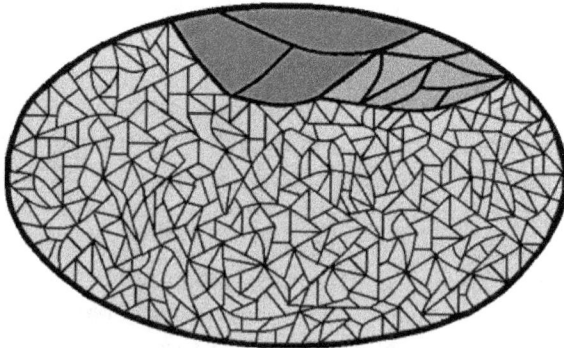

Figure 2.6: Uneven Mix with Too Many Small Parts Yielding "Small Is Exceedingly Beautiful"

big constitutes only approximately 10% of the area, and the medium only about 10% as well. Portions such as 40%, 50%, or even 60% are much more typical for the small in the physical world, while the 80% depicted here is an exaggeration that does not really happen often. Here the small is by far more numerous than the big, over and above the configuration of Figure 2.5.

Figure 2.7 depicts an unnatural quantitative configuration that rarely occurs in the physical world. Here the big constitutes approximately 85% of the entire area (i.e. entire quantity), while the small constitutes only approximately 5% of the area, and the medium only about 10%. Here the big managed to be more numerous than the small as well as more numerous than the medium by strongly dominating overall area at the expense of the small and the medium.

Apart from any *static* mathematical arguments, one could actually postulate the equitable mix of all sizes via heuristic argument involving the existence of random consolidation and fragmentation *dynamic* forces acting upon the quantities. These forces are thought of as physical, chemical, geological, or biological forces, as opposed to any abstract mathematical model of quantitative consolidations and fragmentations. The mechanism driving the system in the direction towards size-equilibrium is the differentiated intensity that these consolidations and fragmentations tendencies occur which depends on the current size configuration of the pieces. The bigger the pieces the stronger is the tendency towards fragmentation. The smaller the pieces the stronger is the

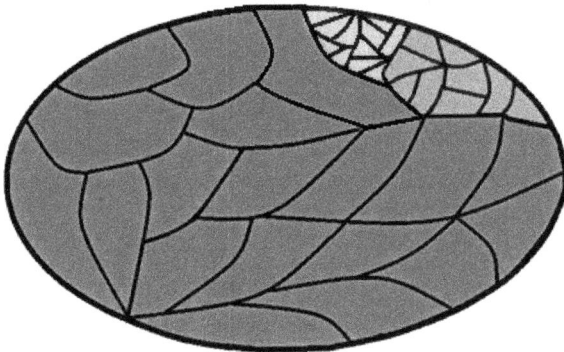

Figure 2.7: Uneven Mix with Too Many Big Parts Yielding Unnatural and Rare Configuration

tendency towards consolidation. Hence a situation with too many big quantities as in Figure 2.7 is unstable, as nature tends to break the many existing big pieces into smaller ones, much more so than it tends to consolidate the very few existing small pieces into bigger ones, thereby gradually changing the composition of sizes in favor of the small. In a similar fashion, a situation with too many small quantities as in Figure 2.6 is unstable, as nature tends to consolidate the many existing small pieces into bigger ones, much more so than it tends to break up the very few existing big pieces into smaller ones, thereby gradually changing the composition of sizes in favor of the big. The equitable and unbiased mix of Figure 2.5 where all 3 sizes share about 1/3 of the entire territory (quantity) is stable. Perhaps such equitable mix of all sizes is the long-term quantitative configuration to which many physical systems tend.

A hypothetical example can be given in the political arena of ancient times. A huge empire is hard to control. Communication and transportation are not well-developed, and so it's difficult for the ruling center in the capital to obtain information about occurrences on the peripheries, and even harder to transport soldiers and material there. Local commanders appointed to rule those far-flung regions are tempted to claim independence and take full control themselves. Hence as a general principle, the larger the empire the greater are the chances and the forces acting upon it towards dissolution and fragmentation. On the other hand, for very small principalities or municipalities ruled by brutal warlords, scheming princes, petty dictators, and such, the opposite forces of conquest, mergers, and consolidations are normally very strong. Here the fear of being absorbed by the well-known proximate neighboring rulers who could easily move armies and material through these short distances is overwhelming and constant. The smaller the political territory the fewer the defenses it can master and the more insecure it feels, and thus the greater the tendency towards conspiracies, attacks, and absorptions. All this leads to plans of preemptive attacks by all sides in order to avoid surprises and conquest, and at times simply due to greed and the desire to rule and exploit larger territory. Such tense state of affairs inevitably ends up with just one successful strongman absorbing his neighboring rivals and ruling all the proximate territory, thus achieving as a consequence much greater political stability. Long-term tendencies work against having many big empires or too many small principalities. Thus a happy balance on the political map between the small and the big is achieved, ensuring that this configuration of the territorial sizes is steady and durable.

Surely, not many entities in the world are derived from actual or physical partition processes.

Yet, if we substitute the words **Composition, Constitution,** or **Consolidation** for **Partition**, and think of the entity represented by the data set under consideration as something composed of many parts, then the partitioning vista and its consequences could be appropriate and applicable for some real-life cases. The three oval-shape entities of Figures 2.5, 2.6, and 2.7 and the conclusions from the related analysis are valid whether it is believed that the entities are actively going to be physically partitioned along the lines inside, or that they are passively composed of all these inner parts — and these two points of view are equivalent as far as resultant quantitative configuration is concern. None of the examples presented in Section 1 can be qualified strictly as partition, not even the example of the atomic analysis of chocolate which is outright composition. The chocolate is composed mostly of the very small hydrogen atoms and of medium-size atoms such as carbon, nitrogen, and oxygen atoms, as well as very few and rare big atoms. The piece of chocolate was certainly not some continuous chunk of primordial matter initially, only to be 'partitioned' later and sorted into its constituents atoms, yet, allowing such fictitious description of the piece of chocolate leads to the same quantitative and size configuration! Clearly, allowing the interpretation of partition as composition enlarges the scope of the analysis.

In other words, applications of partition models in the natural sciences do not have to explicitly assume actual or physical fragmentations of the whole into parts for the results outlined in this chapter to hold. Indeed, a natural entity at times can be thought of as being composed of much smaller parts, or that it exists as the consolidation of numerous separate and smaller parts held together by some physical force or because of any other reason, leading to the same conclusion regarding quantitative and size configuration as for actual partition.

In truth, the exact three categories for sizes, namely Small, Medium, and Big shown in Figures 2.5, 2.6, and 2.7 rarely occur as such in nature; they are artificial and arbitrary drawing for the sake of demonstrating the principles involved regarding partitions. Typically, Mother Nature cannot be as exact, orderly, and regimented when randomly partitioning an overall quantity into parts by sticking only to three sizes throughout the entire process. Yet, the neat arrangements of partitions into parts with only three possible sizes in the above figures are presented for pedagogical purposes, in order to gain broad insight about the different possibilities of partitions, and in order to inform on the fact regarding which generic configuration is more common and which is relatively rare in real-life data and partitions. There exist two extreme poles above and below such exact and neat partitioning into three sizes as in the above figures, outlined in the following two opposing scenarios:

(i) A partition into identical parts, utilizing only one size for all the parts, as shown approximately in Figure 2.8 (difficulties in artistically drawing the exact same size for all areas led to slight variations). The small is beautiful phenomenon cannot be manifested here since there is only one size involved. Certainly, Mother Nature would almost never partition in such even and controlled manner, except in some exceedingly rare occasions and cases.

(ii) A partition into totally distinct parts, utilizing as many sizes as there are parts, as shown in Figure 2.9. This is the most common scenario in random partition processes having the flexibility of producing parts of any real, fractional, or integral values

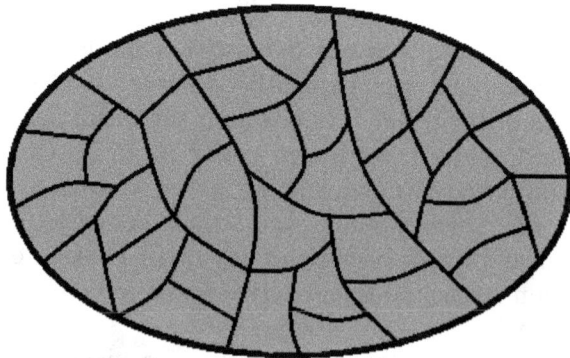

Figure 2.8:   Partitioning a Quantity into Equal Parts Utilizing One Size Only

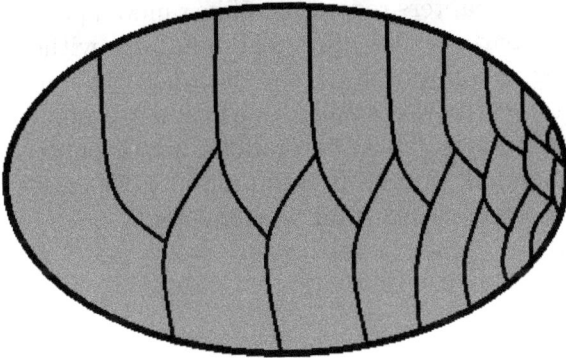

Figure 2.9:   Partitioning a Quantity into Totally Distinct Parts

whatsoever. True randomness in typical partition processes ensures that Mother Nature would produce her parts in a thoroughly chaotic manner, and without any repetition of any single quantity/size whatsoever.

Yet, guaranteeing that the resultant set of parts contains distinct parts without any repetition of values does not immediately imply the small is beautiful phenomenon, rather a careful examination of the resultant set of parts must be performed. Surely the fanatical and stubborn statistician can superficially manufacture the phenomenon by generously defining Small on a wide range endowing it many parts, while restricting the definition of Big to a narrower range and thus depriving it of the ability to earn many parts. And surely, another statistician with opposite tendencies, a contrarian, might place the threshold or cutoff boundary points between Small, Medium, and Big in such a way as to greatly benefit Big, letting it earn the majority of parts. Yet, as discussed in chapter 24 ("Proper Criteria for Sizes Balancing the Chances of Small and Big") proper definition of sizes requires that we assign equal sub-interval length to all sizes; and that the entire range is divided fairly into equal territorial segments for the various sizes. Each oval-shape type of partition as in Figure 2.9 with totally distinct parts requires careful definition of sizes according to the entire range, as well as subsequent careful count of sizes according to these size definitions — in order to determine whether or not the small is beautiful phenomenon manifests itself there.

The next two chapters regarding refine random partition processes (allowing the parts to be any real numbers without restricting them to integral values) illustrate the main driving force behind the manifestation of the small is beautiful phenomenon for their resultant sets of parts. These two random partition processes will be computer-simulated, and their resultant sets of parts will be shown to exhibit the small is beautiful phenomenon.

Chapter 29

Random Dependent Partition Is Always in Favor of the Small

The generic small is beautiful principle found in Integer Partitions is applicable to fractional partitions as well, and this fact lends the principle much wider scope and nearly universal applicability in almost all partition schemes. One particular partition scheme in this context is of a well-structured arrangement of repeatedly partitioning a single quantity randomly into many parts, and this is coined as "Random Dependent Partition". This process is best exemplified by randomly breaking a big rock in multiple stages into much smaller pieces, and this example of the generic idea is coined as "Random Rock Breaking". Surely the description of a rock only serves as a vivid example, and the generic idea here is the repeated break up of a single quantity into smaller and smaller values, culminating in the final set of much smaller values.

In spite of the liberal use of the adjective "random", the process of Random Dependent Partition actually follows a strict partition procedure with exact and carefully executed stages. In the first stage the rock is broken into two pieces using a random pair of percentage values, such as 23% and 77% for example. Then the second stage starts with the orderly breaking of each of the two pieces in a random fashion using two new random pairs of percentage values, resulting in four pieces altogether. In the third stage, each of the four pieces is broken into two pieces using four new random pairs of percentage values, resulting in eight pieces altogether, and so forth.

An essential feature leading rapidly to the small is beautiful phenomenon here is the random manner by which each piece is broken into two smaller pieces, namely that the pair of percentage values are always chosen randomly anew at each stage and for each piece. It is helpful to envision an especially-manufactured roulette in a respectable and honest casino. The roulette wheel contains 99 pockets of numbers and one thrown ball which randomly falls into one of the pockets. These 99 values are marked as $\{1\%, 2\%, 3\%, \ldots, 97\%, 98\%, 99\%\}$ so that all possible (integral) percentage values can be obtained by chance. For example, if the ball lands inside the 25% pocket, then the 75%–25% pair of percentages are employed to break the next piece of rock. Figure 2.10 depicts the physical arrangement of such roulette in one imaginary casino.

Let us illustrate Random Dependent Partition with one concrete example of a six-stage process.

A **500** kilogram rock is broken in the first stage via the random 18%–82% percentage pair, yielding two new pieces $[500] \times [18/100] =$ **90** and $[500] \times [82/100] =$ **410**.

In the second stage of the process, firstly, the 90 piece is broken via the random 71%–29% percentage pair, yielding two new pieces $[90.0] \times [71/100] =$ **63.9** and $[90.0] \times [29/100] =$ **26.1**.

Secondly, the 410 piece is broken via the random 86%–14% percentage pair, yielding two new pieces $[410] \times [86/100] =$ **352.6** and $[410] \times [14/100] =$ **57.4**.

In the third stage, the four random pairs of percentage values for the four pieces above about to be broken are respectively: 72%–28%, 29%–71%, 38%–62%, 52%–48%.

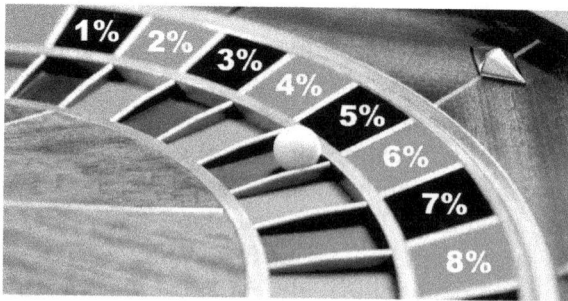

Figure 2.10: Roulette with 99 Pockets Determining the Percent Breakup

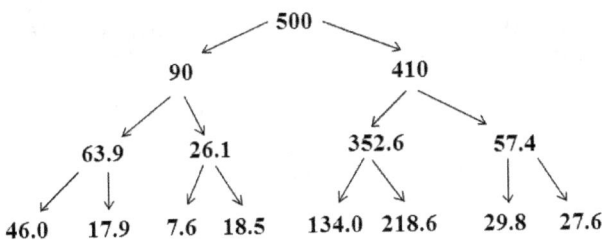

Figure 2.11: Random 500-Kilogram Rock Breaking in Three Stages

This leads to eight new pieces $\{46.0, 17.9, 7.6, 18.5, 134.0,$
$218.6, 29.8, 27.6\}$.

Figure 2.11 depicts the process from the original 500-kilogram
rock to the end of the third stage.

In the fourth stage, these eight pieces are broken in a similar
fashion, leading to 16 new pieces [*percentage values are not shown
for the sake of brevity*]:

8.3	37.7	2.5	15.4	5.1	2.5	15.8	2.8	61.6	72.4	54.7
164.0	17.6	12.2	2.2	25.3						

In the fifth stage, these 16 pieces are broken in a similar fashion,
leading to 32 new pieces:

3.8	4.5	17.7	20.0	0.3	2.2	14.9	0.5	0.2	4.9
1.2	1.3	11.2	4.6	2.4	0.3	30.2	31.4	61.5	10.9
45.9	8.7	113.1	50.8	4.2	13.4	5.1	7.1	0.5	1.7
22.6	2.8								

In the sixth stage, these 32 pieces are broken in a similar fashion,
leading to 64 new pieces:

1.87	1.94	3.00	1.48	15.25	2.48	4.00	16.00
0.27	0.06	1.00	1.18	2.69	12.24	0.14	0.32
0.07	0.08	1.52	3.39	0.23	0.99	1.07	0.20
10.18	1.01	2.65	1.92	2.10	0.34	0.24	0.10
12.68	17.52	18.55	12.89	19.68	41.82	6.62	4.23
35.35	10.56	3.76	4.98	66.75	46.38	9.15	41.68
2.96	1.27	10.44	2.94	2.36	2.78	6.81	0.28
0.48	0.00	1.63	0.09	13.08	9.48	0.86	1.92

The final set of the 64 pieces after the sixth stage, sorted low to high, is as follows:

0.005	0.06	0.07	0.08	0.09	0.10	0.14	0.20
0.23	0.24	0.27	0.28	0.32	0.34	0.48	0.86
0.99	1.00	1.01	1.07	1.18	1.27	1.48	1.52
1.63	1.87	1.92	1.92	1.94	2.10	2.36	2.48
2.65	2.69	2.78	2.94	2.96	3.00	3.39	3.76
4.00	4.23	4.98	6.62	6.81	9.15	9.48	10.18
10.44	10.56	12.24	12.68	12.89	13.08	15.25	16.00
17.52	18.55	19.68	35.35	41.68	41.82	46.38	66.75

A cursory look at the above set of numbers roughly confirms the small is beautiful feature of the pieces, and this observation is decisively confirmed by the detailed histogram of Figure 2.12. It should be noted that the first three bins in the histogram are of 1-kilogram width, while the last four bins are of 5-kilogram width. There are five pieces weighing over 23 kilograms which are not shown in the histogram as they are very sparsely and thinly spread far out on the horizontal axis, rendering the big there even rarer.

Skeptical or cynical readers might suspect the author for possible manipulations of the roulette simulations so as to arrive at many

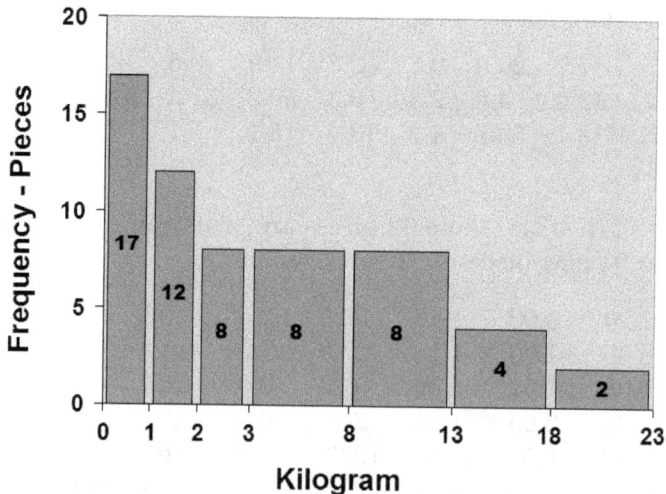

Figure 2.12: Histogram of Pieces After the Sixth Stage — Random 500-Kilogram Rock Breaking

small pieces but only few big pieces. Yet the above results could be easily reproduced and verified with the aid of a standard personal computer. MS-Excel for example provides a random number generator called RAND() which yields random values uniformly and evenly distributed between 0 and 1, such as 0.964125, and 0.176387, and so forth. These random fractions could be interpreted here as the random percentage values. Indeed, this more general usage of fractions from the **real number line on (0, 1)** — without restricting random percentages to integral values — is what constitutes part of the setup and the real definition of Random Dependent Partition.

There are two fundamental differences between the small is beautiful model of Integer Partition as outlined in the previous chapter, and this model of Random Dependent Partition. Firstly, Integer Partition model strictly and exclusively considers only integral values, while Random Dependent Partition liberally includes any fractional and real ones. Secondly, in Integer Partition the model incorporates or aggregates all possible partition scenarios into one vast data set, while in Random Dependent Partition there is no need to aggregate distinct partition scenarios, and instead only a single trajectory or scenario of the breakup of the original rock is considered.

The general *dependency* of the weights of the pieces at each stage upon the weights of the pieces in the previous stage should be noted carefully. Surely, this dependency on the previous stage also involves dependency on random elements, namely the random percentage values which determine how to break the pieces of the previous stage. Hence the process concocts new values at each stage out of the old values of the previous stage, mixing in random elements as well.

Chapter 30

Random Real Partition Is Always in Favor of the Small

A random and spontaneous partitioning process lacking any dependent stages leads to the small is beautiful phenomenon just as well. Here the phenomenon is encountered without any constraints, free of strict sequential procedures, encompassing integral as well as fractional values. Instead of breaking the original quantity in sequential stages, a comprehensive plan of how and where to cut or break is contemplated beforehand, and then at an opportune and appropriate grand moment the partition is fully executed, cutting or breaking simultaneously in all the planned places and locations. The process is coined as "Random Real Partition". This process is best exemplified by the cutting of one-dimensional long pipe at random locations along its length, and this example of the generic idea is coined as "Random Pipe Breaking".

Let us provide a concrete numerical example. A 15-meter-long metal pipe is to be randomly partitioned into 30 parts. This is accomplished by obtaining 29 independent random points along the pipe — prior to the moment of the actual cutting — to serve as marks indicating where the pipe should be partitioned. These marks constitute the grand plan of the entire partition process, and they are generated via independent simulations from the continuous **Uniform(0, 15)**.

Actually, distances from the left edge of the pipe are obtained via 29 computer simulations using the random number generator

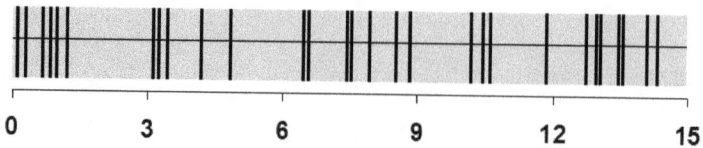

Figure 2.13: The 29 Marks Randomly Scratched along the 15-Meter Pipe

uniformly distributed between 0 and 1. These 29 simulated numbers then provide random percentage values to determine the positions of the marks along the 15-meter length pipe, for example $(0.28) \times (15) = 4.2$, or $(0.71) \times (15) = 10.7$, and so forth.

Following the simulation of each such random position, a mark on the physical pipe is made accordingly with a black marker or by gently scratching the location with a saw to indicate where to do the actual cutting later when the grand moment of execution arrives. It is only at the end of the long sequence of 29 simulations and markings that actual cutting and sawing at these marked locations take place, breaking the long pipe into 30 parts of (usually) totally distinct lengths.

Figure 2.13 depicts the actual 29 marks along the pipe obtained via computer simulations.

The 29 random values generating the marks on the pipe, order from low to high; including the left edge of 0.000 as extra, and the right edge of 15.000 as extra, are outlined as follows:

0.000	0.140	0.301	0.710	0.868	1.003	1.243	3.163
3.273	3.453	4.181	4.871	6.500	6.606	7.453	7.550
7.962	8.555	8.847	10.207	10.464	11.652	11.882	12.750
12.947	13.060	13.452	13.564	14.123	14.376	15.000	

Calculating distances between the marks (i.e. differences) by subtracting from each value greater than zero its adjacent value on the left, we obtain the set of the lengths of the parts of the pipe:

0.140	0.161	0.409	0.158	0.135	0.239	1.921	0.109	0.180
0.728	0.689	1.629	0.106	0.846	0.097	0.412	0.593	0.292
1.361	0.257	1.188	0.230	0.868	0.197	0.113	0.392	0.112
0.559	0.253	0.624						

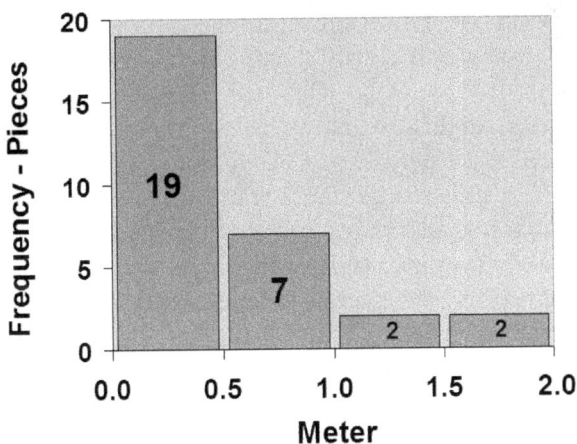

Figure 2.14: Histogram of Resultant 30 Parts in Random Real Partition

Ordering these distances low to high, we obtain the final ordered set of the lengths of the parts:

0.097	0.106	0.109	0.112	0.113	0.135	0.140	0.158	0.161
0.180	0.197	0.230	0.239	0.253	0.257	0.292	0.392	0.409
0.412	0.559	0.593	0.624	0.689	0.728	0.846	0.868	1.188
1.361	1.629	1.921						

Clearly, the small is beautiful principle is evidently valid for this set of 30 values. Had this pipe been cut deterministically and fairly into 30 equal parts, then each part would have been $15/30 = 0.5$ meter long, hence the value of 0.5 serves as the benchmark for the "truly middle size". In contrast, for this random partition here, the majority of the parts are shorter than 0.5 meter; fewer are longer than 0.5 meter; and all this is surely in the spirit of the small is beautiful principle.

Figure 2.14 depicts the histogram of these 30 parts after partition, confirming visually the manifestation of the small is beautiful phenomenon for this random partition process.

The process of Random Real Partition involves random and totally independent numerical choices for the partition of a given quantity. Typical examples of the process involve the cutting of one-dimensional long "pipe" at random locations along its length — giving each section and each part of the pipe equal probability in being marked by using the continuous Uniform. Such pipe examples fit naturally with the defined process since it is very easy and straightforward to mark those locations along the one-dimensional pipe beforehand as planned partition. In contrast, "rock" is three-dimensional, and marking surfaces and areas or presenting diagrams and maps for cutting beforehand is extremely cumbersome and complex, and totally impractical.

The term "real" in the phrase "Random Real Partition" refers to the fact that there is no restriction on having only integral values as was the case for Integer Partition, and that the parts or the pieces can attain any fractional, rational, real, and irrational value. In addition, the positions of the marks along the one-dimensional pipe as in Figure 2.13 invoke the concept of the "real line", or the "real number line" in mathematics, which is the line whose points are the set R of all real numbers (also viewed as a geometric or Euclidean space).

Chapter 31

Numerous and Distinct Parts in Partitions are Necessary Conditions

In order to arrive at the resultant small is beautiful quantitative configuration in random partitions, it is necessary at a minimum that the process thoroughly breaks the original quantity into a set of numerous and mostly distinct smaller parts. Some numerical examples follow:

Breaking $\{15\}$ into $\{5,5,5\}$ or into $\{1,2,3,3,3,3\}$ involve repeated values such as 5 or 3, and thus these partition processes are not conducive to the small is beautiful phenomenon. In addition, these partitions yield very few parts, and this is also not conducive to the phenomenon.

Breaking $\{15\}$ into $\{1,2,3,4,5\}$ is a bit more conducive to the phenomenon since the parts are of distinct quantities without any repetition of values, yet since this process partitions 15 into only 5 smaller parts, it is rendered insufficient to lead to the phenomenon.

Breaking $\{15\}$ into the set of 30 parts as in the example of pipe breaking of the previous chapter yields the following set of parts (sorted from low to high):

$\{0.097,\ 0.106,\ 0.109,\ 0.112,\ 0.113,\ 0.135,\ 0.140,\ 0.158,\ 0.161,$
$0.180,\ 0.197,\ 0.230,\ 0.239,\ 0.253,\ 0.257,\ 0.292,\ 0.392,\ 0.409,$
$0.412,\ 0.559,\ 0.593,\ 0.624,\ 0.689,\ 0.728,\ 0.846,\ 0.868,\ 1.188,$
$1.361,\ 1.629,\ 1.921\}$

This pipe partition process is quite conducive to the phenomenon for two reasons. Firstly, the process produces numerous parts, namely thirty. Secondly, the resultant set of parts is of totally distinct quantities, without any repetition of values whatsoever.

Certainly, Random Dependent Partition (Rock) with only 1, 2, or 3 stages say, producing only 2, 4, or 8 pieces, does not result in the small is beautiful quantitative configuration, due to scarcity in the number of resultant pieces. In the same vein, Random Real Partition (Pipe) with only 2 to 8 parts say, does not result in the small is beautiful quantitative configuration, due to scarcity in the number of resultant parts. For these two partition processes to arrive at the small is beautiful quantitative configuration, it is necessary that enough randomness has been executed throughout the system, and that the system has produced plenty of parts.

Breaking {15} into the set of 30 parts, with 15 parts of value 0.40, 10 parts of value 0.50, and only 5 parts of value 0.80, is a partition process which does indeed produce numerous parts.

$$\{0.40,\ 0.40,\ 0.40,\ 0.40,\ 0.40,\ 0.40,\ 0.40,\ 0.40,\ 0.40,\ 0.40,$$
$$0.40,\ 0.40,\ 0.40,\ 0.40,\ 0.40,\ 0.50,\ 0.50,\ 0.50,\ 0.50,\ 0.50,$$
$$0.50,\ 0.50,\ 0.50,\ 0.50,\ 0.50,\ 0.80,\ 0.80,\ 0.80,\ 0.80,\ 0.80\}$$

The resultant set of parts is typically not of distinct values, since there are frequent repetitions here, nonetheless, this partition process has been carefully and deliberately (and artificially) calibrated to produce more small parts than big parts, hence it manifests the small is beautiful phenomenon. Surely, it must be acknowledged that this meticulous partition process was not random in any way, but rather deliberately produced and planned.

Breaking {15} into the set of 30 parts, with 10 parts of value 0.20, 10 parts of value 0.50, and 10 parts of value 0.80, is a partition process which does indeed produce numerous parts.

$$\{0.20,\ 0.20,\ 0.20,\ 0.20,\ 0.20,\ 0.20,\ 0.20,\ 0.20,\ 0.20,\ 0.20,$$
$$0.50,\ 0.50,\ 0.50,\ 0.50,\ 0.50,\ 0.50,\ 0.50,\ 0.50,\ 0.50,\ 0.50,$$
$$0.80,\ 0.80,\ 0.80,\ 0.80,\ 0.80,\ 0.80,\ 0.80,\ 0.80,\ 0.80,\ 0.80\}$$

Yet, the resultant set of parts is typically not of distinct values, since there are frequent repetitions of values. This partition process is definitely not in the spirit of the phenomenon, since all 3 sizes are of equal [10 times] frequency. Surely, it must be acknowledged that also this meticulous partition process was not random in any way, but rather deliberately produced and planned.

Since rock breaking and pipe breaking are performed randomly, and since these partition processes are not limited to integral values but are rather performed on the basis of all possible real numbers, there should normally be no repetition, and the chances of finding many/most/all parts having identical values are exceedingly small, and formally the probability for repeated real values is zero. Given that computer simulations are thorough and refined, then not even a single repetition of parts should be found in these random partition processes. Randomness ensures that almost nothing is steady and equal in the final resultant quantitative configuration of the set of parts after partition.

Chapter 32

Conclusion: Random Partitions and the Small Is Beautiful Phenomenon

In Mathematical Statistics it is shown that distances between random markings along the length of one spacial dimension are distributed as in the Exponential Distribution. This distribution is strictly positively skewed, monotonically falling to the right, and having a strong preference for the small. This result implies that Random Real Partition exhibits the small is beautiful phenomenon. Detailed discussion about the connection between the Exponential Distribution and Random Real Partition, as well as one concrete example of Random Pipe Breaking fitting nicely into this distribution shall be given at the end of this section.

Surely there are in principle many random partition processes other than Random Dependent Partition and Random Real Partition. These two processes are perhaps the most straightforward or natural ways to break a quantity into parts in a random fashion. The concluding general statement regarding the manifestation of the small is beautiful phenomenon in random partition processes is based on very consistent and broad empirical testing and evidences, as well as sound conceptual underpinning. In part it also rests on some rigorous mathematical results.

The general statement is based on three assumptions:

(i) Given that partition is performed on the real number basis and not exclusively on integers.
(ii) Given that partition is truly random.
(iii) Given that partition thoroughly breaks the original quantity into numerous refined parts.

Then all the (numerous) parts are of distinct values, and the small is beautiful quantitative configuration is found in the resultant set of parts.

Chapter 33

Multiplication Processes Lead to the Small Is Beautiful Phenomenon

Almost all random or deterministic multiplication processes induce a dramatic increase in skewness where the small becomes relatively numerous and the big becomes relatively rare. Surely, the single product of the multiplication of only two numbers is not to be considered here, since the resultant single product is neither small nor big, but rather it's just itself, and there are no competing sizes here to consider. Instead, a set of N numbers called A is to be multiplied by another set of M numbers called B. The phrase "multiplication of two sets of numbers" implies that each and every number in set A is to be multiplied by each and every number of set B, producing N × M products. In other words, all possible multiplications between the two sets are attempted, and the entirety of these products constitutes the newly created set of numbers, called A × B. For example:

Set A = $\{8, 3, 5\}$

Set B = $\{11, 47, 26\}$

A × B = $\{(8) \times (11), (8) \times (47), (8) \times (26), (3) \times (11), (3) \times (47),$
$(3) \times (26), (5) \times (11), (5) \times (47), (5) \times (26)\}$

A × B = $\{88, 376, 208, 33, 141, 78, 55, 235, 130\}$

Let us examine the quantities within the 10 by 10 multiplication table that we all were forced to memorize in our elementary school years. In this example, the intrinsic characteristics of multiplication processes with regard to resultant sizes shall be explored and compared. Such an analysis is done at the most primitive and basic level, at the arithmetic and quantitative level, before going on to the rigorous mathematical level. The quest is to start out with this very particular example, and then perhaps if possible, lend the conclusions derived from this case universality and applicability in almost all multiplication processes.

For the 10 by 10 multiplication table, the entire range from 1 to 100 is to be partitioned into 10 quantitative sections of equal 10-unit width each, namely $[1, 10]$, $[11, 20]$, $[21, 30]$, $[31, 40]$, $[41, 50]$, $[51, 60]$, $[61, 70]$, $[71, 80]$, $[81, 90]$, $[91, 100]$, in order to count the values falling within each section. The goal is to group the 100 products of the multiplication table according to sizes. Figure 2.15 depicts this quantitative partitioning arrangement of the entire multiplicative territory by size. Figure 2.16 depicts the histogram of the values falling within each section. Surprisingly, a decisive trend regarding the occurrences of products within the sections is found here. The section of the smallest quantities (1 to 10) has 27 values falling

X	1	2	3	4	5	6	7	8	9	10
1	1	2	3	4	5	6	7	8	9	10
2	2	4	6	8	10	12	14	16	18	20
3	3	6	9	12	15	18	21	24	27	30
4	4	8	12	16	20	24	28	32	36	40
5	5	10	15	20	25	30	35	40	45	50
6	6	12	18	24	30	36	42	48	54	60
7	7	14	21	28	35	42	49	56	63	70
8	8	16	24	32	40	48	56	64	72	80
9	9	18	27	36	45	54	63	72	81	90
10	10	20	30	40	50	60	70	80	90	100

Figure 2.15: Quantitative Territorial Partitioning of the 10 by 10 Multiplication Table

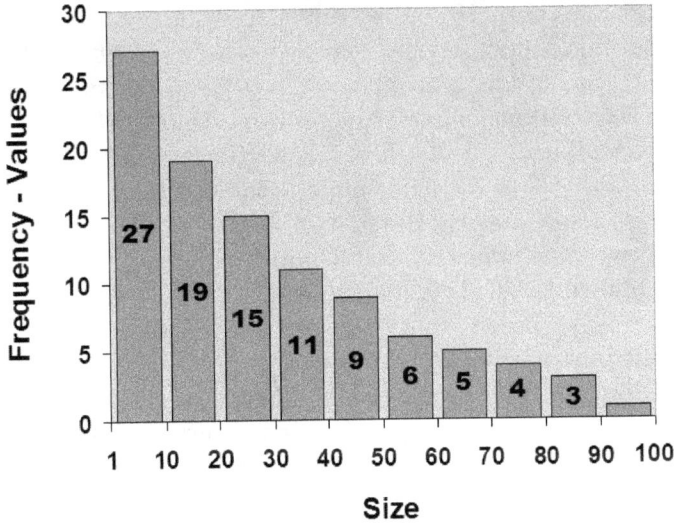

Figure 2.16: The 10 by 10 Multiplication Histogram — Small Is Beautiful

within it; while the section of the biggest quantities (91 to 100) has only 1 value falling within it.

The sequence of all the values falling within these 10 sections, and presented in order according to sizes is: $\{27, 19, 15, 11, 9, 6, 5, 4, 3, 1\}$. Clearly, the sections pertaining to bigger quantities have less values falling within them, as the count of values monotonically decreases. The small is definitely more numerous than the big in our standard multiplication table. The crucial lesson learnt from this multiplication process is that this tendency may be very general, and that it should be present in many other multiplication processes, and not only for our particular 10 by 10 multiplication table! There is nothing unique about our standard multiplication table or the numbers from 1 to 10, and therefore we wish to extrapolate this result to almost all other multiplication processes.

The results of Figures 2.15 and 2.16 can also be interpreted as a particular casino game where two dice having 10 faces each are thrown, and the values on the two faces are multiplied by each other. Clearly the above result nicely explains why the casino was almost always winning the Four-Dice Multiplication Game of Chapter 22. Our 10 by 10 multiplication table can also be interpreted in a more formal sense in the context of theoretical statistics as the random multiplicative process of two *discrete* random Uniform Distributions, namely the product: Uniform$\{1, 2, 3, 4, 5, 6, 7, 8, 9, 10\}$ × Uniform$\{1, 2, 3, 4, 5, 6, 7, 8, 9, 10\}$, instead of regarding it just as a useful tool of the deterministic table of multiplication. More generally, it could also be thought of as the random multiplicative process of two *continuous* random Uniform Distributions, namely the product Uniform[1, 11) × Uniform[1, 11).

It should be noted that for the above result about the 10 by 10 multiplication table, the starting point is the uniform and even distribution of the original numbers about to be multiplied, namely $\{1, 2, 3, 4, 5, 6, 7, 8, 9, 10\}$, where neither small nor big is more numerous than the other. Yet, from such even distribution, we arrived (via multiplications) at a decisively skewed and uneven distribution where the small is more numerous than the big.

Exception to the rule is found only in multiplications of rare sets of numbers with almost no variability, namely data sets having extremely low order of magnitude. This case shall be discussed in details later in this book.

In Mathematical Statistics, the random product of numerous independent realizations from an identical random variable is known to be the Lognormal Distribution in the limit as the number of these multiplied realizations becomes large enough. This seminal result which is derived from the Multiplicative Central Limit Theorem (MCLT) has very lax requirements and only a few restrictions, ensuring broad applicability for almost all types of multiplied variables (i.e. for almost all multiplication processes). Most significantly, the restriction on having an identical distribution can often be relaxed assuming other easily obtained conditions, thus MCLT can usually be applied to the random product of several distinct random variables. Since the Lognormal Distribution is structured in the small is beautiful quantitative configuration [almost always, that is, for shape parameters above 1.0 roughly], MCLT generally implies that most multiplication processes are strongly in favor of the small. Even though MCLT is a stated as a limiting process, the overall skewed quantitative configuration of the small is beautiful phenomenon is encountered every early on, well before the distribution truly attains strong resembles to the Lognormal

Distribution, and thus the applicability of MCLT in our context is much broader than merely achieving the Lognormal configuration itself.

The scope of applications here is truly enormous, encompassing almost all disciplines. The *customary multiplicative form* of the vast majority of the equations in physics, chemistry, astronomy, economics, biology, engineering and other disciplines, as well as the numerous expressions relating to their applications and specific results, leads to the manifestation of the small is beautiful phenomenon in the natural sciences and real-life data. These data sets are almost always with high enough variability (i.e. high order of magnitude) and thus they almost never suffer from the above-mentioned exception to the rule.

As discussed in Kossovsky (2014) Chapter 90, Isaac Newton gave us $F = M \times A$, not $F = M + A$. Hence the derivations and results due to the expression Force = Mass × Acceleration are related to multiplication processes, not to addition processes. Newton also gave us the law of universal gravitation $F_G = G \times M_1 \times M_2/R^2$ which is written in multiplicative and divisional forms, and not in additive and subtractive style such as say $F_G = G + M_1 + M_2 - R^2$.

Chapter 34

Exponential Growth Series and the Small Is Beautiful Phenomenon

Exponential growth occurs when the growth of a given quantity is proportional to the quantity's current value, so that the larger the current value, the more it grows. As an example, if the growth of the population of rabbits in the wild is 5% per year, given stable environmental conditions in general, such as food availability, the spread of predator population in the vicinity, weather, and the normal level of microbial threat, then a population of say 1,000 rabbits will bring into the world 50 newborn rabbits that year. Several years later, when the rabbit population reaches 100,000, and under the same environmental conditions, they will bring into the world 5,000 newborn rabbits that year. It stands to reason that the more rabbits there are in the fields, the more baby rabbits are born there each year. In the same vein, as newly born stars form and grow, gathering intergalactic gas and particles before achieving critical mass and start burning, the bigger the star the more gravitational force it exerts on its surroundings, the more mass it attracts, and the more it grows — per any given time period. The advance from one value to the next; from one period (say end of 2014) to the next time period (say end of 2015); assuming an exponential 5% growth per period, for example, can be expressed arithmetically as

$$\begin{aligned} \text{VALUE}_{2015} &= \text{VALUE}_{2014} + 5\% \text{ of VALUE}_{2014} \\ &= [1] \times \text{VALUE}_{2014} + [5/100] \times \text{VALUE}_{2014} \\ &= [1 + 5/100] \times \text{VALUE}_{2014} \\ &= [1.05] \times \text{VALUE}_{2014} \\ &= [\text{Growth Factor}] \times \text{VALUE}_{2014} \end{aligned}$$

Let us consider for example 5% exponential growth series starting from the base of 500.0 in the initial year. By the second year it will grow to $[500.0] \times [1.05] = 525.0$. By the third year it will grow to $[525.0] \times [1.05] = 551.3$. By the fourth year it will grow to $[551.3] \times [1.05] = 578.8$, and so forth. It should be noted carefully that the term for the fourth year could also be expressed as repeated multiplications, namely as $[500.0] \times [1.05] \times [1.05] \times [1.05] = 578.8$.

Clearly, exponential growth series are intimately connected to multiplication processes! Not surprisingly, the small is beautiful phenomenon is found in the exponential growth series as well.

Let us examine the exponential 5% growth series from 500.00 to 9339.59. This involves 60 growth periods, measured in years perhaps. A young 20-year-old person is imagined to invest for his or her future late retirement by depositing \$500 in a special bank account which will be frozen for 60 years, while earning 5% [compound] interest per year, and where no withdrawals or deposits are allowed. The intention is to have some funds available when reaching the ripe old age of 80. The entire series (rounded to zero decimal place and thus excluding the fractional parts) is:

500	525	551	579	608	638	670	704	739
776	814	855	898	943	990	1039	1091	1146
1203	1263	1327	1393	1463	1536	1613	1693	1778
1867	1960	2058	2161	2269	2382	2502	2627	2758
2896	3041	3193	3352	3520	3696	3881	4075	4279
4493	4717	4953	5201	5461	5734	6020	6321	6637
6969	7318	7684	8068	8471	8895	9340		

The small could be beautiful here only if we incorporate and examine numerous, some, or perhaps even just several consecutive readings of the series. The consideration of the last element only (the value in the last year) couldn't lead to any size comparison since there is only one size and only one number to examine.

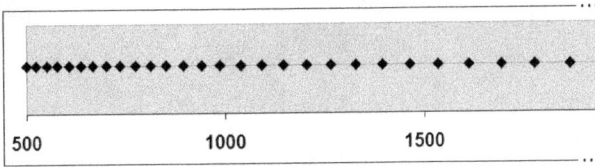

Figure 2.17: Higher Concentration for Small and Diluted Spread for Big

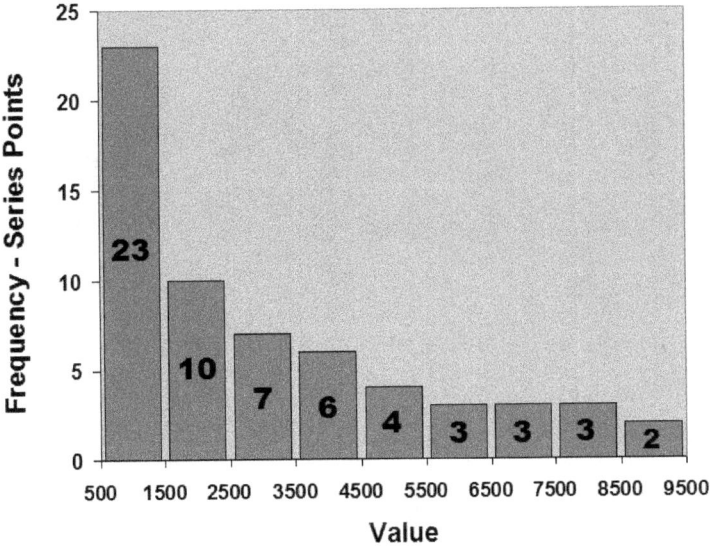

Figure 2.18: Values in 5% Exponential Growth — Absent Time Dimension

Figure 2.17 depicts the accelerated march of the 5% exponential growth series from the base value of 500 up to around 2000. Clearly, there are more values falling near the smaller value of 500 than around the bigger value of 1000. And there are more values falling near 1000 than around the bigger value of 1500. The series consistency thins out and gets diluted as we move to bigger values on the right.

Figure 2.18 depicts the histogram of the entire exponential growth series on the wider range from 500 to 9500, in even steps of 1000, disregarding the time dimension altogether. The small is clearly by far more numerous than the big almost everywhere along the horizontal axis.

Chapter 35

Explanation of the Small Is Beautiful Phenomenon in Growth Series

Clearly, if one wishes to focus on the addition aspect of repeated multiplications in exponential growth series such as in the example of the previous chapter, then it is noted that progressively bigger and bigger values are being added. A factor of 1.05 for the 5% exponential growth series implies adding only about 25 in the beginning near 500; adding about 50 near 1000, and adding a whopping 500 approximately at the end near 10,000. This explains why values are accelerating as the series marches forward, as seen clearly in Figure 2.17.

Let us prove in another way that distances between consecutive elements in exponential growth series are constantly expanding — regardless of the specific rate of growth — and thus that values thin out and get diluted as we move forward to bigger values on the right. This in turn would prove that the histogram is skewed; falling on the right, and that small is beautiful. In Figure 2.19, the symbol F represents the multiplicative factor per period, assuming a particular percent of growth, namely F = (1 + percent/100). For example, for 5% exponential growth series, F = (1 + 5/100) = 1.05. Growth here is assumed to be real and positive (unless exponential decay is considered), therefore F must be greater than 1. The symbol B represents the base value, namely the initial value in the first year before any growth takes place.

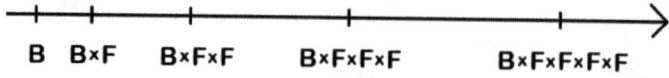

Figure 2.19:   The Accelerated March of Exponential Growth

At the first year: B

At the second year: B × F

At the third year: B × F × F

At the fourth year: B × F × F × F

At the fifth year: B × F × F × F × F

   Let us evaluate the distances between the successive elements of Figure 2.19 as follows:

$$BF - B = B(F - 1)$$
$$BFF - BF = B(F - 1)F$$
$$BFFF - BFF = B(F - 1)FF$$
$$BFFFF - BFFF = B(F - 1)FFF$$

   Evidently, distances between elements are increasing by a factor of F, and which is larger than 1. The distances are $B(F-1)F^N$ and with $F > 1$, implying that distances between each consecutive pair keep growing, and that therefore the concentration of values is constantly decreasing.

Chapter 36

The Fibonacci Series and the Small Is Beautiful Phenomenon

The "Fibonacci series" is a case which relates indirectly to exponential growth series, and where the small is beautiful phenomenon manifests itself decisively. The first 21 elements are $\{1, 1, 2, 3, 5, 8, 13, 21, 34, 55, 89, 144, 233, 377, 610, 987, 1597, 2584, 4181, 6765, 10946\}$, and even superficial glance at these numbers confirms that the small vastly outnumbers the big. The series arbitrarily starts with the two initial elements of $\{1, 1\}$. Subsequent elements beginning with the third value are calculated deterministically as the addition of the previous two elements, namely as $X_N = X_{N-1} + X_{N-2}$.

Surprisingly, even though the series is defined additively, it quickly turns into a multiplicative process, with the Golden Ratio 1.618034 being the multiplicative factor. For example, the 15th element is 610 and it is approximately equal to the Golden Ratio times the 14th element 377, namely $377 \times 1.618034 = 609.99881 \approx 610$. Hence, very early on, from approximately the 11th element, the series starts to behave as a standard exponential 61.8034% growth series, and the expression $X_{NEXT} = 1.618034 \times X_{PREVIOUS}$ can be used quite accurately for all subsequent elements in place of the original additive expression.

Figure 2.20 depicts the first 10 elements of the Fibonacci series. Clearly, the small is more numerous than the big, and the series gets diluted and thins out on the right of the x-axis as bigger values are considered.

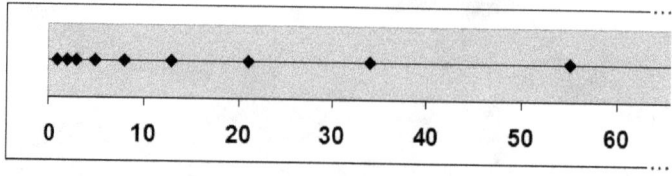

Figure 2.20:   The Accelerated March of the Fibonacci — Small Is Beautiful

Chapter 37

Growth Model for Planets, Stars, Cities, and Bank Accounts

As mentioned earlier, no discussion about relative frequencies can take place regarding the last element of a single exponential growth series, as there is only one size and only one number to consider. On the other hand, we may consider the last element in a large collection of a variety of exponential growth series with distinct initial base values, distinct percent growth rates, and distinct numbers of growth period. Not surprisingly perhaps, the small is found to vastly outnumber the big in such collections of final values of random growth series. This innovative idea of focusing on the last element of a large collection of exponential growth series with random parameters was suggested and analyzed by the mathematician Kenneth Ross.

Imagined the nation of Australia in the south-western Pacific being established and developed after the European discovery of the continent in 1606. Ignoring — at our own moral peril — all the prior towns and settlements of the indigenous Aboriginal Australians, we could then assume a series of towns and cities being established independently at different times, by different numbers of initial pioneers, growing at different rates, and with inter-migration excluded in order to simplify the analysis. Here we allow for fractional values of persons, since the model represents the generic idea of quantitative growth, and not necessarily only of integral values.

In summarizing the model, the country is said to have cities of varying and random age chosen uniformly between 1 and 410 years;

some are very old and established cities, some are not quite old, and some are relatively new cities. In addition, for each city, the initial base value representing the number of pioneers who established the city is randomly and uniformly chosen from 1 to 50; and the growth rate is randomly and uniformly chosen from 1.0% to 3.0% as well (and that singular selected rate is then considered as fixed and applied to all the growth years for the given city). The current year is set to 2016. The variable under consideration is the current snapshot of the populations of all existing cities and towns in Australia after $(2016 - 1606) = 410$ years, and considering only the last elements of the growth series in the current year 2016, but not including historical population records throughout the years.

The table in Figure 2.21 depicts one computer simulation run with 30 such imaginary cities. The table is sorted from low to high by current population count. For example, the city at the bottom of the table with the fictitious name Sydnia is 374 years old. It was established in 1642 by a wild and mischievous band of 42 ex-convicts. It grew constantly by 2.3% growth rate per year. The current (civilized and well-behaved) population of Sydnia is calculated as:

$$(42) \times (1.023) \times (1.023) \ldots 374 \text{ times} \ldots (1.023) \times (1.023)$$
$$= (42) \times (1.023^{374}) = 203,397$$

The histogram of the population snapshot in Figure 2.22 clearly shows that small cities are much more numerous than big cities in this simulation run. The first bin is of population from 1 to 1,000. The second bin is of population from 1,000 to 10,000, and thus much wider. The last two bins are even wider. The choices for bin widths are arbitrary in a sense, chosen in order to arrive at the best visual presentation of this very small data set. For economical and brevity purposes, the histogram is not drawn exactly according to the authentic ratios or scales of these distinct bin widths. In any case, the big — in spite of its huge bin-width advantage — is much less frequent than the small.

A close scrutiny of the three simulated parameters in Figure 2.21 would confirm that they are approximately uniformly distributed, and without any bias towards the small. And yet, the small managed to triumph decisively in the calculated variable of the current population count!

City's Name	Year Established	City's Age	Growth Rate	Number of Pioneers	Current Population
Tasmotia	1989	27	2.6%	6	11
Orangia	1990	26	2.4%	7	13
Tamworthia	1988	28	2.8%	9	19
Geraldtonia	2014	2	1.8%	25	26
Launcestonia	2008	8	1.6%	46	52
Devonportia	1783	233	1.4%	2	54
Coffsia	1961	55	2.9%	27	132
Gatheftia	1961	55	2.6%	41	171
Waggatia	1904	112	2.7%	10	192
Mackayia	1847	169	1.0%	36	197
Rockhamia	1831	185	1.1%	29	204
Herveynia	1954	62	3.0%	34	209
Bundabergia	1846	170	2.3%	5	232
Dullotia	1914	102	2.5%	32	396
Bunburyia	1848	168	1.7%	42	665
Cookia	1768	248	1.7%	15	1,050
Gussotonia	1900	116	2.8%	42	1,103
Wollongongia	1859	157	2.2%	46	1,501
Hobartia	1820	196	2.4%	15	1,685
Geelongia	1681	335	1.4%	19	2,319
Townsvillia	1767	249	1.9%	24	2,363
Cairnsia	1702	314	1.6%	46	6,412
Darwinia	1630	386	1.6%	15	7,216
Busseltonia	1778	238	2.8%	20	14,998
Bathurstia	1624	392	1.7%	32	20,811
Canberrnia	1657	359	2.1%	37	55,133
Pertha	1760	256	3.0%	39	73,134
Brisbania	1701	315	2.7%	27	103,401
Melbournia	1710	306	2.8%	23	114,268
Sydnia	1642	374	2.3%	42	203,397

Figure 2.21: Exponential Growth Model for Fictitious Cities in Australia

Kenneth Ross has suggested another more realistic mathematical model where the growth rate for each city varies randomly anew for each distinct growth period. Such a model requires for each city multiple simulations for growth rates for all the periods. Such a model employs more randomness and uncertainty in the system and as a result it naturally converges even more rapidly and easily to the quantitative configuration of the small is beautiful phenomenon.

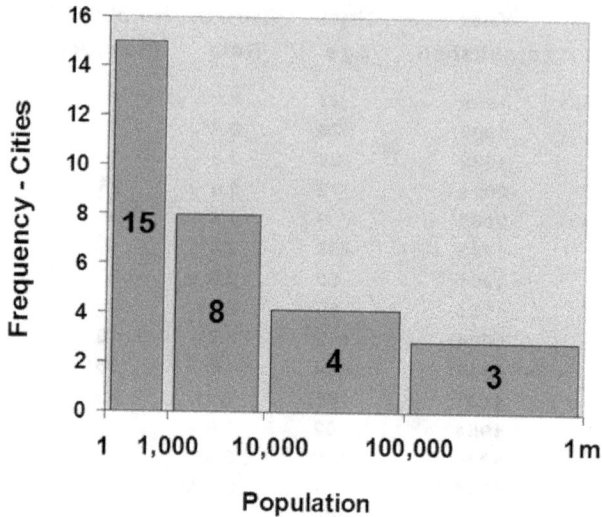

Figure 2.22:   The Small Outnumbers the Big in Simulated Model for Cities

The above result could be re-produced and verified via computer simulations using other choices for the range of the parameters of the three Uniform variables, leading to the same quantitative conclusions, and especially so if many more cities are simulated instead of merely 30 cities as in this example. The driving force towards the small is beautiful phenomenon in this multi-growth model is the multiplicative form in how the generic quantity under consideration is increasing.

The enormous significance of this result regarding the final elements of multiple growth series is that it can serve as a generic quantitative model for all entities that spring into being gradually via random growth, be it a set of cities and towns with growing populations, rivers forming and enlarging gradually along an incredibly slow geological time scale, planet and star formations, the growth of a huge number of biological cells within an organism, and so forth. The potential scope covered by the result of this chapter is enormous!

As for another example, Prudent Retirement Bank specializes exclusively in long-term 15-year, 20-year, 25-year, and 30-year savings accounts. The accounts earn fixed interest payments each month at an interest rate determined by the treasury and bond markets on the day of the opening of the account. Customers are not permitted to make any withdrawals or deposits before the ending date. Rules regarding initial amount invested are quite flexible, but there are $1,000 minimum and $5,000 maximum restrictions. The statistician or data analyst at the bank is asked by the manager to take a quantitative snapshot of all the accounts as one vast data set as of the end of a particular day. Surely the data analyst will conclude that accounts with very big amounts are rare, and that the vast majority of them are with small amounts. Here for the bank, accounts are growing in approximately the same random style as cities in fictitious Australia are growing. The accounts are growing at distinct interest rates; having a variety of initial amounts invested; and are of different duration or ages.

Our home, the Milky Way Galaxy, is an average-sized spiral galaxy, and astronomers estimate that it contains up to 400 billion stars of various sizes and brightness. Stars born and die, constantly. The death of very big stars is accompanied by spectacular explosions, such as supernovas which can be visible even in far away galaxies. These cataclysmic explosions may trigger the formation of new stars, continuing the cycle of stellar birth and death. At any given point in (cosmological) time, the galaxy contains stars with a huge variety of ages. Some are exceedingly old and some are relatively very young. Star's longevity depends only on how much mass it has. Big heavy stars that are say ten times the mass of the sun last ("only") about 100,000,000 years. Average-size stars such as our sun last about 13,000,000,000 years. Small stars that are say one-tenth the mass of the sun are blessed with incredible longevity, and could last about 100,000,000,000 years or longer! The author's short life expectancy of merely 78 years (if any) pales in comparison!

Such state of galactic affairs is a fertile ground for our multi-growth model. It could be assumed that there is a wide range of star growth rates and that it varies sufficiently, depending on the local density of dust and particles. Surely in space rich with primordial material, growth is faster, and vice versa. Hence in terms of our model, the growing entities (stars) are of distinct growth rates, of distinct ages [regarding growth time before maturing into their sizes], and presumably of distinct "initial base mass". Consequently, it stands to reason that current observations of sizes of stars should most certainly manifest the small is beautiful phenomenon. Indeed, actual data on stars decisively support this principle as discussed in Chapter 8, as does the study on exoplanets in Chapter 6.

Chapter 38

Data Aggregation Leads to the Small Is Beautiful Phenomenon

It may not be obvious, but surprisingly quite often real-life data sets consist of numerous, smaller, and "more elemental" mini sub-sets. Therefore, a given data set which appears to exist independently as a whole, may actually be made of several data components which are aggregated in order to arrive at that larger set of data. The implication of such aggregations to relative quantities and sizes is profound, since results are almost always in favor of the small. Indeed, it can be demonstrated in general that appending various data sets into a singular and much larger data set leads to the small is beautiful phenomenon whenever these data sets commonly start from a very low value, ideally such as 0 or 1, and terminate at highly differentiated endpoints, so that some span short intervals while others span longer intervals. Let us demonstrate this quantitative tendency in data aggregation by combining the following six imaginary data sets:

Data Set A: $\{2, 3, 5, 7\}$

Data Set B: $\{1, 4, 6, 9, 13, 14\}$

Data Set C: $\{2, 6, 7, 9, 11, 15, 16, 21\}$

Data Set D: $\{1, 2, 6, 8, 13, 14, 19, 23, 25\}$

Data Set E: $\{3, 4, 8, 12, 15, 19, 22, 24, 29, 35, 41\}$

Data Set F: $\{1, 5, 8, 11, 12, 17, 19, 24, 27, 32, 38, 43, 47\}$

Individually, each data set does not show any quantitative preference for the small, yet when these six data sets are merged together to become that aggregated and large data set, the resultant quantitative configuration is decisively in favor of the small and biased against the big.

The combined data set A, B, C, D, E, F:

{2, 3, 5, 7, 1, 4, 6, 9, 13, 14, 2, 6, 7, 9, 11, 15, 16, 21, 1, 2, 6, 8, 13, 14, 19, 23, 25, 3, 4, 8, 12, 15, 19, 22, 24, 29, 35, 41, 1, 5, 8, 11, 12, 17, 19, 24, 27, 32, 38, 43, 47}

The combined data set A, B, C, D, E, F, ordered from low to high:

{1, 1, 1, 2, 2, 2, 3, 3, 4, 4, 5, 5, 6, 6, 6, 7, 7, 8, 8, 8, 9, 9, 11, 11, 12, 12, 13, 13, 14, 14, 15, 15, 16, 17, 19, 19, 19, 21, 22, 23, 24, 24, 25, 27, 29, 32, 35, 38, 41, 43, 47}

It should be noted that we are aggregating the numbers residing within the data sets, but we are not aggregating the data sets in and of themselves. The focus here is on all the numbers, not on the data sets. This means that each number is assigned equal weight and proportion, yet not each data set earns equal weight and proportion. For example, post-aggregation, set A has only 4 values within the aggregated data, while set F has 13 values, hence set F is relatively more represented in the aggregated data set. As for a more concrete example: had set F contained 13 primes exclusively, while set A only 4 composites exclusively, then the aggregated data set would have been strongly swayed by set F, containing more primes than composites.

Figure 2.23 depicts the histogram of these 51 values. In spite of the fact that each component data set is quantitatively structured in an even and balanced manner approximately [very roughly as in the Uniform Distribution], where no size is significantly preferred over any other size, yet for the aggregated data set the small decisively outnumbers the big.

The dynamics behind such tendency to produce numerous small quantities but only few big ones is the differentiated overlapping of ranges for the aggregated data set. Overlapping here occurs more on the left for small values, and less so on the right for big values.

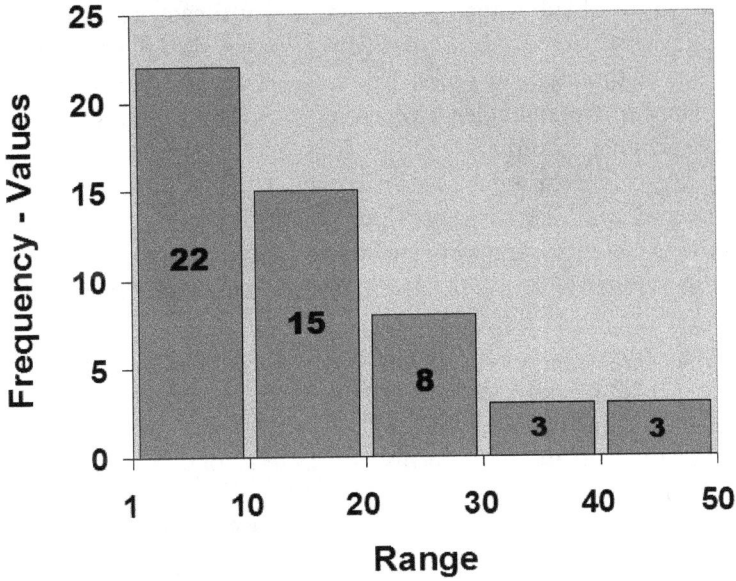

Figure 2.23: The Aggregation of Six Data Sets — Small Is Beautiful

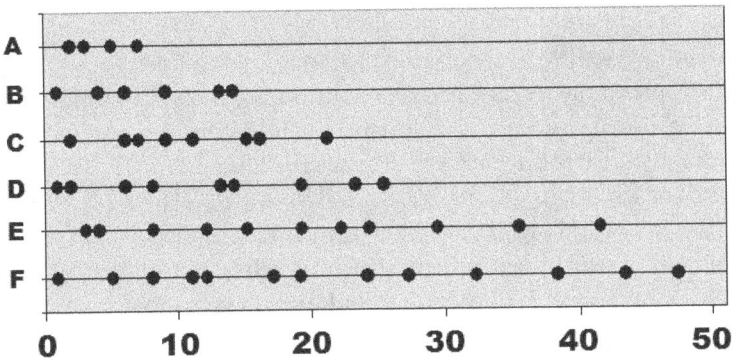

Figure 2.24: Visualizing the Piling up of More Values on the Left

Figure 2.24 depicts these six data sets superimposed, thus allowing us to visualize the intense piling up of numerous small values that occurs on the left, in sharp contrast to the diluted piling up of only few big numbers that occurs on the right.

It should be noted that some types of data aggregations do not lead to any such differentiated piling up of values in favor of the small on the left side of the range. The existence of these exceptions here to the principle of the small is beautiful is quite significant and should always be acknowledged so as to avoid overgeneralization and rush to judgment in the consideration of data aggregations. There are two types of extreme counter examples. The first type is data aggregation without any overlapping whatsoever, where all the ranges are totally disconnected and distinct, so that no piling up occurs at all. The second type is data aggregation where all the ranges span almost the same territory, and overlapping is so consistent and repetitive, so that nothing changes quantitatively.

Aggregating $\{2, 6, 8\}$, $\{9, 11, 23\}$, $\{27, 31, 38, 41, 53\}$, $\{57, 61, 64, 72\}$, $\{75, 82, 88, 96\}$ into a singular data set provides an example for the first type of exception, where component data sets do not overlap at all, and instead simply continue to expand forward on the overall range.

Aggregating $\{1, 4, 11\}$, $\{1, 3, 7, 9\}$, $\{2, 4, 5, 8, 10\}$, $\{1, 4, 6, 7, 11\}$, $\{3, 5, 7, 10\}$ into a singular data set provides an example for the second type of exception, where all component data sets overlap each other over approximately the same range of values; where all component data sets begin and terminate at approximately the same points.

Another possible (although very rare) exception to the small is beautiful principle is an inverted differentiation in overlapping of ranges, where more values pile up on the right for big values, and less values pile up on the left for small values. As an example for such rare cases, aggregating $\{1, 6, 10, 13, 14, 18\}$, $\{5, 9, 11, 17, 20, 21\}$, $\{3, 8, 13, 15, 19\}$, $\{7, 13, 16, 17, 20\}$ into a singular data set results in an inverted quantitative configuration, with many big values and fewer small values; where the big happened to be beautiful (i.e. negative skewness).

Yet, the small is beautiful principle (i.e. positive skewness) is still valid in most cases of data aggregations, because the original setup of A to F data sets of Figure 2.24 is the most natural and typical case in the aggregations of real-life physical and abstract data sets. The three other alternatives are artificial and imaginary for the most part, and even though they might hold true in some very particular circumstances, they are simply quite rare. For the vast majority of real-life data creation and data aggregation, the natural process is of measurements that start out from 0 or 1 if not from another very low value, while terminating at a wide variety of distinct bigger values. This fact of life is stated without any concrete proof or argument, and without any experiment or empirical analysis of data. Instead, the author is referring to common sense and general intuition. For those with a lot of experience in real-life data, the above assertion should appear quite natural and compelling as it probably correlates with their knowledge of how data is typically formed and compiled. Surely, most physical and real-life data is exclusively of positive values, and that implies that 0 is the lower limit that all such component data sets must share. Event data, such as accidents per day or per month, births per year in a particular municipality, and so forth, are all exclusively of positive integral values, and that implies that 1 is the lower limit that all such component data sets must share. The upper limit on the other hand is typically unique to the particular component data set under consideration, and thus it varies considerably in the entire scheme of data aggregation.

Surely, not all data sets are derived from aggregation of smaller data components. For example, data on star sizes is not derived from any supposed aggregation. Instead, individual stars are observed and examined one by one, and the value of each star is added to the already collected large set of star values. For data sets derived from data aggregation, this could occur either in an implicit way or in an explicit way. A good example of implicit data aggregation is address data where the focus of the data analyst is solely on the house number. Here address data for an entire city can be implicitly or indirectly modeled as the aggregation of all the mini data sets of all its existing streets. Let us illustrate this more vividly by listing house numbers for several streets pertaining to a data set of a post office branch in one particular small town, and which typically may be listed as follows:

$\{1, 2, 3, 4, 5, 6\}$ — Floral Drive

$\{1, 2, 3, 4, 5, 6, 7, 8, 9, 10, 11, 12, 13, 14, 15\}$ — Pine Avenue

$\{1, 2, 3, 4, 5, 6, 7, 8, 9, 10, 11, 12, 13, 14, 15, 16, 17, 18, 19\}$ – Main Street

$\{1, 2, 3, 4, 5, 6, 7, 8, 9, 10, 11, 12\}$ — South Street

$\{1, 2, 3, 4, 5, 6, 7, 8, 9\}$ — Lodge Street

All streets necessarily start at house number 1, namely the first house on the left or right corner, but each street terminates at different house number depending on the length of the street, hence many numbers pile up on the left range of small values, leaving the right range of big values diluted and sparse. The aggregation of all house numbers in all the streets in a given city constitutes an implicit data aggregation where the small is decisively beautiful.

The empirical study on address data pertaining to Prince Edward Island in Eastern Canada of Chapter 15 where small house numbers decisively outnumber big house numbers confirms the above theoretical and conceptual arguments.

Chapter 39

Three-Dice Selection Game as Data Aggregation or Probability Scheme

Could the three-dice selection game of Chapter 23 be viewed simply as data aggregation?

Given that all three dice (four-sided, eight-sided, twelve-sided) occur with equal chance of 33.33%, the game can be thought of as the aggregation of all possible dice occurrences of $\{1, 2, 3, 4\}$, $\{1, 2, 3, 4, 5, 6, 7, 8\}$, and $\{1, 2, 3, 4, 5, 6, 7, 8, 9, 10, 11, 12\}$. Certainly, all dice necessarily start at 1, but they terminate at a variety of maximum values, namely at 4, 8, and 12, depending on the die. Figure 2.25 depicts all possible occurrences of the game as data aggregation.

Figure 2.26 depicts the simple-minded construction of a histogram of 24 such three-dice selection games. The focus of the histogram in Figure 2.26 is on the *possibilities*, and it is based on Figure 2.25, naively taking that data aggregation interpretation of the dice game all too seriously, totally ignoring *probabilities*. The histogram is constructed via a simple count of the points within Figure 2.25, and where $\{1, 2, 3, 4\}$ sums up to 12, $\{5, 6, 7, 8\}$ sums up to 8, and $\{9, 10, 11, 12\}$ sums up to 4. Since $(12 + 8 + 4) = 24$, the histogram is based on 24 such three-dice selection games. However, while the drawing of Figure 2.25 is perfectly correct in the sense that it serves as a good visualization tool, the histogram of Figure 2.26 is misguided and it needs some adjustments, as it lacks probability analysis.

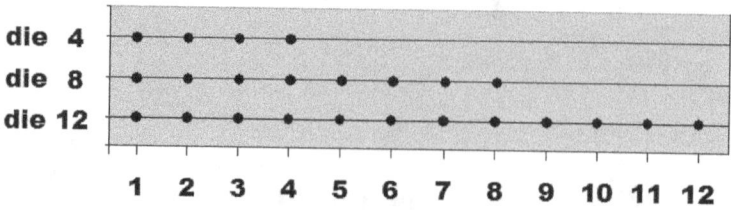

Figure 2.25: Visualization of How More Values Pile Up on the Left

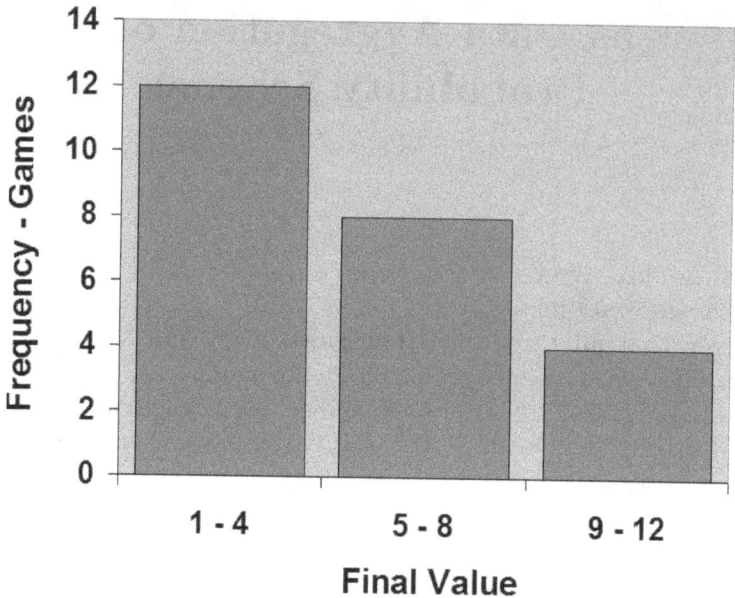

Figure 2.26: Misguided Histogram Construction of 24 Three-Dice Games

The three-dice selection game differs somehow from data aggregation in the sense that the focus here is on the data sets (dice), where each die earns an equal 33.33% probability, as opposed to focusing on the numbers (dice faces) themselves. Therefore, even though the twelve-sided die has three times more faces than the faces in the four-sided die, it still earns equal weight of only 33.33% within the entire scheme. This implies that not all numbers (dice faces) come with equal probabilities. In order to uphold the 33.33% per die rule, is it essential that for each die we distribute that 33.33% portion equally to all its members, namely to divide that 33.3% portion per die among all its faces, die by die, therefore:

Each face in the four-sided die is assigned $(33.33\%)/(4) = 8.333\%$ chance ~8.3% chance.

Each face in the eight-sided die is assigned $(33.33\%)/(8) = 4.166\%$ chance ~4.2% chance.

Each face in the twelve-sided die is assigned $(33.33\%)/(12) = 2.777\%$ chance ~2.8% chance.

Figure 2.27 depicts the probability for each particular occurrence in the three-dice selection game (the percent sign "%" is omitted for brevity). It should be noted that adding percentages along each row in Figure 2.27 yields 33.3%. For example: 8.3%+8.3%+8.3%+8.3% = 33.3%. Adding all the percentages in the entire table of Figure 2.27 yields 100%.

It follows that the chance of obtaining 1 as the final value in the game is calculated by summing the chance that the twelve-sided die yields 1, plus the chance that the eight-sided die yields 1, plus the chance that the four-sided die yields 1. This leads to 2.77%+4.16%+ 8.33% = **15.3%**. This calculation also applies equally to the chance of obtaining 2, 3, or 4, as the final value. Occurrences of final values 5, 6, 7, or 8, do not involve the four-sided die in any way, hence the calculation for each such final value is 2.77% + 4.16% = **6.9%**. Occurrences of final values 9, 10, 11, or 12, involve only the 12-sided die, hence the chance for each such final value is **2.8%**.

The stage is now set for the proper construction of the theoretical histogram which would be based on the probabilities for each final

die 4	8.3	8.3	8.3	8.3								
die 8	4.2	4.2	4.2	4.2	4.2	4.2	4.2	4.2				
die 12	2.8	2.8	2.8	2.8	2.8	2.8	2.8	2.8	2.8	2.8	2.8	2.8
	1	2	3	4	5	6	7	8	9	10	11	12

Figure 2.27: Probabilities for Each Possible Occurrence in Three-Dice Selection Game

value from 1 to 12, namely on the following distribution:

1	2	3	4	5	6	7	8	9	10	11	12
15.3%	15.3%	15.3%	15.3%	6.9%	6.9%	6.9%	6.9%	2.8%	2.8%	2.8%	2.8%

As in the misguided Figure 2.26 which uses three bins, each of 4-unit width, the newly constructed proper histogram of probabilities would also be of three bins. Let us add probabilities as follows:

For the first bin of $\{1, 2, 3, 4\}$ we add $(15.3\%+15.3\%+15.3\%+15.3) =$ **61.2%** probability.

For the second bin of $\{5, 6, 7, 8\}$ we add $(6.9\%+6.9\%+6.9\%+6.9\%) =$ **27.6%** probability.

For the third bin of $\{9, 10, 11, 12\}$ we add $(2.8\% + 2.8\% + 2.8\% + 2.8\%) =$ **11.2%** probability.

Since the histogram in Figure 2.26 contains 24 games in total, the newly constructed proper histogram would be similarly based on the theoretical chanced proportions of occurrences for 24 hypothetical games about to be played, instead of showing pure percentages. Hence the above set of four probabilities shall be converted into proportions of 24 games by multiplying them by 24 as follows:

$0.612 \times 24 = 14.7$

$0.276 \times 24 = 6.6$

$0.112 \times 24 = 2.7$

Based on the above calculations, the proper histogram for the three-dice selection game is depicted in Figure 2.28. More precisely, this histogram depicts the expected result in 24 such three-dice selection games. It is noted that this proper histogram of Figure 2.28 is skewer than the misguided histogram of Figure 2.26.

Computer simulations with 20,000 runs of the three-dice selection game yielded results which were highly compatible with the theoretical distribution, strongly corroborating the above discussion. These results are as follows:

1	2	3	4	5	6	7	8	9	10	11	12
15.6%	15.4%	15.5%	14.9%	6.9%	6.7%	6.9%	6.8%	2.8%	2.9%	2.9%	2.7%

Surely, the results of only 100 games simulated in Chapter 23 are not sufficiently numerous in a statistical sense, and therefore these

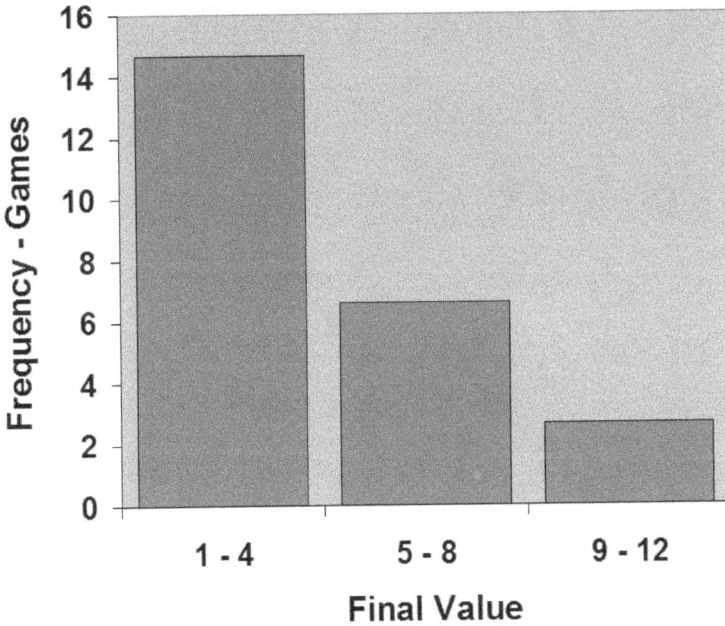

Figure 2.28: Proper Histogram Construction of the Expectation of 24 Three-Dice Games

results cannot be as close to the theoretical probabilities as the above simulations with 20,000 runs. Nonetheless they do come a bit close to the theoretical expectation, as follows:

1	2	3	4	5	6	7	8	9	10	11	12
22.0%	16.0%	18.0%	9.0%	7.0%	7.0%	4.0/%	5.0%	4.0%	3.0%	2.0%	3.0%

Chapter 40

Thirty-Dice Selection Game Is Consistently in Favor of the Small

Evidently, the above three-dice selection game involving four-sided, eight-sided, and twelve-sided dice, is numerically not sufficiently smooth, as the probability abruptly decreases from 4 to 5, and from 8 to 9, while remaining constant elsewhere. In order to provide a more instructive example of the quantitative results in multi-dice selection games, another numerically smooth dice selection game shall be examined involving all possible dice from one-sided die to 30-sided die, and all of those dice in between.

That numerically smooth example is given by the casino which has invested in a very large and expensive bronze 30-sided die, as well as in 30 inexpensive and much smaller plastic dice, all having a variety of sides, including the smallest plastic die with only one side constantly showing the value of 1; the two-sided plastic die showing the values 1 and 2, and so forth, up to the small plastic 30-sided die. The rules of this game are similar to the rules of the three-dice selection game, so that the large bronze 30-sided die is thrown first, and the value of the face that comes up determines the type of the smaller plastic die to play for the next throw, leading to the final value in the game. For example, if the large bronze die shows the face of 17, then the small plastic 17-sided die is the next to be thrown, and the outcome of this second throw constitutes the final value. There is a small plastic die to play for each face occurrence of the large bronze die!

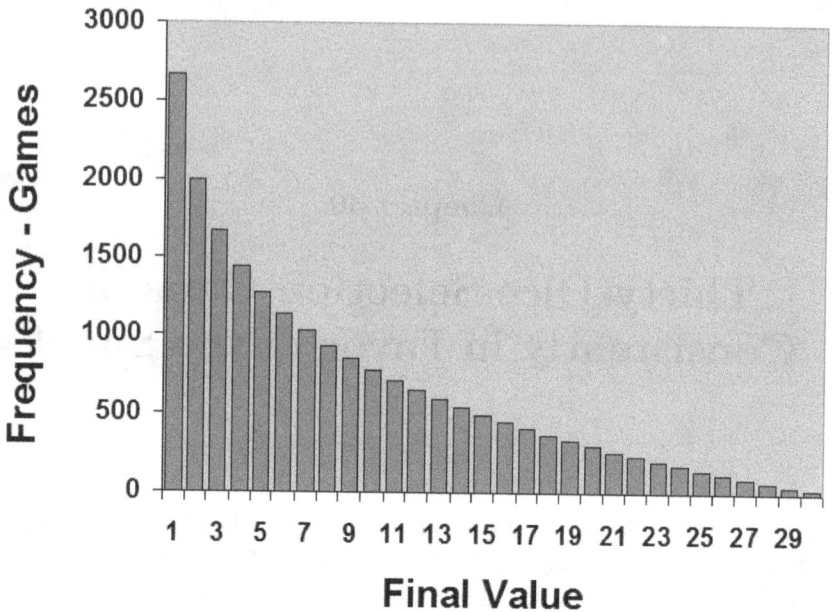

Figure 2.29:   Histogram of Final Value Expectations for 20,000 30-Dice Games

The theoretical calculation for a final value of 30 involves first the probability of obtaining the face of 30 in the large 30-sided bronze die which is 1/30, superimposed with the second probability of obtaining again the face of 30 in the smaller plastic 30-sided die, and which is also 1/30. The probability of this rare event is calculated as $(1/30)(1/30)$ or 0.00111, namely 0.111%.

Theoretical calculation for a final value of 29 involves the summing of the probability that the bronze 30-sided die yields 30 and that the plastic 30-sided die yields 29, plus the probability that the bronze 30-sided die yields 29 and that the plastic 29-sided die yields 29. The probability of this event is calculated as $(1/30)(1/30) + (1/30)(1/29) = 0.00226$, namely 0.226%.

Theoretical calculation for a final value of 1 involves the summing of the probability that the bronze 30-sided die yields 1 and that the plastic 1-sided die yields 1, plus the probability that the bronze 30-sided die yields 2 and that the plastic 2-sided die yields 1, plus the probability that the bronze 30-sided die yields 3 and that the plastic three-sided die yields 1, and so forth for all the other dice, up to and including the probability that the bronze 30-sided die yields 30 and that the plastic 30-sided die yields 1. Hence the probability of a final value 1 in the game is given by the long summation of all of these 30 terms, as follows: $(1/30)(1/1) + (1/30)(1/2) + (1/30)(1/3) + \cdots$ etc. $+(1/30)(1/29) + (1/30)(1/30) = 0.13317$, namely 13.317%.

The histogram of Figure 2.29 depicts the theoretical expectations of the final values for 20,000 such 30-dice selection games. For example, in 20,000 games, the probability of obtaining the final value of 1 is calculated as (Probability of 1) $\times$ (20,000) $=$ (0.13317) $\times$ (20,000) $=$ 2663.3. The histogram falls off smoothly and gently everywhere along the curve, and where the relatively small is always more numerous than its bigger neighboring counterpart on the right everywhere in the histogram.

Chapter 41

Chains of Statistical Distributions and the Small Is Beautiful Phenomenon

In order to generalize dice selection games, such as the three-dice and thirty-dice selection games above, formal notations are introduced. A fair and balanced die with N sides constitutes the Discrete Uniform Distribution, and it is denoted as **DIE(N)**, where N is considered to be the parameter of the distribution, and which must be an integral number. DIE(N) refers to the arrangement where each integer from 1 to N inclusively earns the identical probability value of 1/N. Hence the 30-dice selection games can be presented or expressed more concisely as **DIE_SECONDARY(DIE_PRIMARY(30))**, where the parameter of the final distribution is not fixed as is usually the case, but rather itself being dependent on another Discrete Uniform Distribution. In performing computer simulations, one random realization from DIE(30) is obtained and this fixed number is called R; then another random realization from DIE(R) is obtain, representing the final value. No such simplistic notation can be written for the three-dice selection game though, since it is a bit more complex and irregular. The unique feature here is the dependency of the parameter of a distribution on yet another distribution, as opposed to the orthodox approach in Mathematical Statistics where parameters are always defined as constants.

In the same vein, data aggregation could be similarly viewed as some parametrical dependency scheme. Two features guarantee the quantitative configuration in favor of the small in the context of data aggregation. The first feature is the common beginning of the

177

approximate minimum for all the component data sets around 0 or 1. The second feature is the random or highly varied termination at the end for the maximum, namely that each component data set has its own particular maximum. Such quantitative tendency or mechanism in aggregations of real-life data sets is one of the main causes of the manifestation of the small is beautiful phenomenon in the world. Formalism in Mathematical Statistics draws inspiration from such types of data aggregations and points to an abstract statistical process coined as "a chain of two Uniform Distributions", namely the statistical chain Uniform(min, Uniform(maxA, maxB)), with the constraint that min < maxA < maxB.

The Continuous Uniform Distribution **Uniform(a, b)** constitutes the set of all possible values from a to b (including fractional and irrational numbers) such that all are with the same chance of occurring. In other words, all comparable sub-ranges having the same width are with equal probability. It is useful to think of **a** as the parameter signifying the minimum, and of **b** as the parameter signifying the maximum. The parameters **a** and **b** are fixed, constant, and well-known numbers, and there is nothing fuzzy or random about them.

The histogram of the Uniform Distribution is horizontal and flat, neither rising nor falling, and therefore all sizes are of equal frequency and importance. The small is not more frequent than the big, or vice versa. The big is not more frequent than the medium, or vice versa.

A chain of two Uniform Distributions model in the general spirit of data aggregations is such where the main or principle distribution is with parameter **a** which is fixed at say 0 or 1, and it is called "common minimum"; together with parameter **b** which is considered to be a fuzzy, uncertain, and random number, drawn from another Uniform Distribution with two specific fixed parameters, and expressed as Uniform(lowest maximum, highest maximum). Here the term "lowest maximum" refers to the maximum value in the narrowest component data set, and the term "highest maximum" refers to the maximum value in the widest component data set. The term "common minimum" refers to the same minimum value of all the component data sets. The chain is then:

Uniform(common minimum, Uniform(lowest maximum, highest maximum))

Yet, a chain of two Uniform Distributions differs from data aggregation in one fundamental way, namely assigning equal importance and probability to all component data sets, as opposed to assigning equal importance and probabilities to all the numbers. The focus of the chain of distribution is on the component data sets (i.e. the Uniform Distributions), giving each one equal probability within the entire scheme, while the focus of data aggregation is on the numbers, giving each number equal probability. Hence, in this sense at least, the chain of Uniform Distributions is not exactly the proper or the ideal model for data aggregations, although it could often serve as a very good approximation of it.

As an example of how data aggregations could (approximately) be modeled by chains of distributions, let us consider the combined data set A, B, C, D, E, and F, as shown in Figure 2.24, and let us attempt to crudely represent its resultant distribution by the model of the two-sequence chain **Uniform(1, Uniform(7, 47))**. The phrase "representing a certain numerical entity by some abstract model" refers to the similarity in resultant quantitative configuration between that numerical entity and the model, so that their histograms appear very similar. Unfortunately, the modeling attempt of this chain of distribution fails to give a good fit, and its numerical result is somewhat different from the aggregated data set. There are three distinct reasons why the above aggregated data set cannot be represented well via such a chain.

The first reason for the failure to fit the chain model to the aggregated data is that the chain pertains to all values, integral as well as fractional and irrational numbers, while the data aggregation set contains only integral values.

The second reason is that there are gaps and jumps in the values of the integers of the various data sets. They do not increase nicely and smoothly by one integer at a time, but rather irregularly. In addition, the maximum boundaries are not increasing smoothly by one integer at a time either, but rather they occur with gaps and jumps as well. In addition, not all data sets begin with the same minimum value of 1; only data sets B, D, F do; the rest of the data sets start with either 2 or 3.

The third reason is that such a chain assigns equal probability for Data Set A of $\{2, 3, 5, 7\}$ with only 4 values, as well as for Data Set F of $\{1, 5, 8, 11, 12, 17, 19, 24, 27, 32, 38, 43, 47\}$ with 13 values.

The chain repeats $\{2, 3, 5, 7\}$ approximately three times in order to give Data Set A equal importance overall, thus injecting for example the set $\{2, 3, 5, 7, 2, 3, 5, 7, 2, 3, 5, 7\}$ into the final aggregated data set, in order to accomplish the goal of equivalency in overall importance for all data sets.

On the other hand, chains of distributions could serve as reasonably good models for multi-dice selection games. For example, the chain **Uniform(1, Uniform(2, 31))** is a fairly good model for the 30-dice selection game, albeit without the restriction of integral values and the ending near 31 instead of exactly 30. In performing computer simulations, one random realization from the distribution Uniform(2, 31) is obtained and this fixed number is called Q; then another random realization from the distribution Uniform(1, Q) is obtain, representing the final value. Indeed the histogram of 20,000 simulations from this chain of two Uniform Distributions is depicted in Figure 2.30, demonstrating compatibility to the histogram in Figure 2.29 of the 30-dice game. The values from the above particular simulation runs of the chain range from the minimum of 1.00028 to the maximum of 30.78165. The compatibility between the chain and

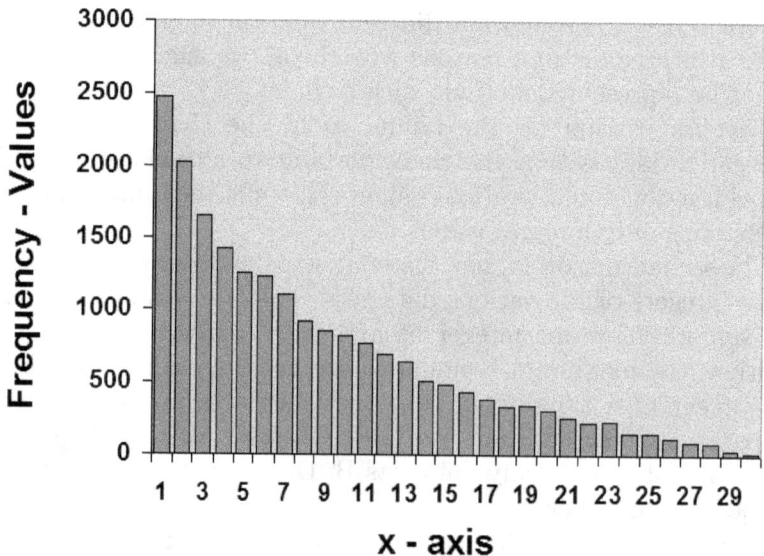

Figure 2.30: Histogram of the Chain Uniform(1, Uniform(2, 31)) Modeling 30-Dice Game

the dice game is derived also from their common focus on either dice sizes or Uniform Distributions, as oppose to the focus on numbers. For example, the 30-dice selection game endows equal weight to die 5 as to die 27, even though die 5 has only 5 faces, while die 27 has 27 faces. In the same vein, the chain above endows equal weight to Uniform(1, 6) as to Uniform(1, 28), even though Uniform(1, 6) has the narrower range of 5, while Uniform(1, 28) has the wider range of 27.

It should be noted that bins in Figure 2.29 of the dice game refer to integers, while bins in Figure 2.30 of the chain refer to ranges between integers. For example, the second bin of the dice game refers to the numbers of games with the final value of integer 2, while the second bin of the chain refers to the numbers of simulation points falling between 2.00 and 3.00.

An obvious application of the chains of distributions to real-life data sets is the house number in address data. Assuming that the shortest street in the town is with six houses, and that the longest street is with 53 houses, as well as the presumed smoothness and approximate uniformity in how lengths of streets are distributed, then the model for house number data in this small town could be given by Uniform(1, Uniform((6 + 1), (53 + 1))), or simply as **Uniform(1, Uniform(7, 54))**. Here all streets necessarily start at house number 1, hence parameter **a** is fixed at 1, but each street terminates at a different house number depending on the length of the street, hence parameter **b** should be variable and derived from yet another Uniform Distribution. Yet, the application of the chain of distributions to house number data suffers from the same issue that challenged data aggregation, namely that short streets, as well as long streets in the chain model, are given equal weight, and this feature of the chain model needs to be adjusted to fit the true configuration of the house number data which assigns more weight to longer streets than to shorter ones. The model also suffers a bit from the discrepancy between the chain which pertains to all values, integral as well as fractional and irrational numbers, and the actual address data which is restricted exclusively to integral values.

Practically almost all types of chains of distributions, applying a variety of distribution forms (not merely the Uniform); with any number of dependent sequences (not merely two); tying up location and/or scale parameter(s), yield quantitative skewness and the small is beautiful phenomenon. There are very few rare exceptions to this rule, such as tying up certain types of shape parameter(s). The discussion in Section 4 about chains of distributions contains more detailed and precise rules about their quantitative behavior. All this lends the chain of distributions drive towards the small is beautiful phenomenon a truly colossal scope of manifestations, occurrences, and applications,

In some cases of chains of distributions, the Normal Distribution serves as the variable parameter. The bell-shaped Normal Distribution is the proper statistical model in several physical measurements such as systolic and diastolic blood pressures, heart rates, blood cholesterol levels, heights of adults, and test scores for many standardized tests. Yet the distribution is not very prevalent in real-life measurements and data sets.

In Mathematical Statistics the distribution is formally denoted as **Normal(μ, σ)**, where the μ (mu) parameter on the left stands for the average, and the σ (sigma) parameter on the right stands for the standard deviation. In the context of relative quantities, the essential feature of the Normal Distribution is that it's symmetrical; hence the small is not more numerous than the big, or vice versa. The medium is by far the most numerous size in Normal Distributions. Here the big and the small are equally less significant, obtaining only mediocre frequencies of occurrence.

Even though individually the Uniform or the Normal Distributions do not exhibit the small is beautiful phenomenon standing alone as a single random variable, and in fact they represent the classic counter examples of the phenomenon, yet when chained together involving some parametrical dependencies, the small is beautiful phenomenon often appears as if out of nowhere!

In one example of a chain of distributions involving both the Uniform and the Normal, the maximum **b** parameter of the Uniform Distribution is chained to the Normal Distribution with mean 17 and standard deviation 4, while the minimum **a** parameter is fixed at 0. More concisely, this chain scheme is expressed as:

Uniform(0, Normal(17, 4)).

Since the minimum is fixed at the low value of 0 while the maximum varies by way of the Normal Distribution, such a scheme strongly resembles data aggregation, hence the expectation here is

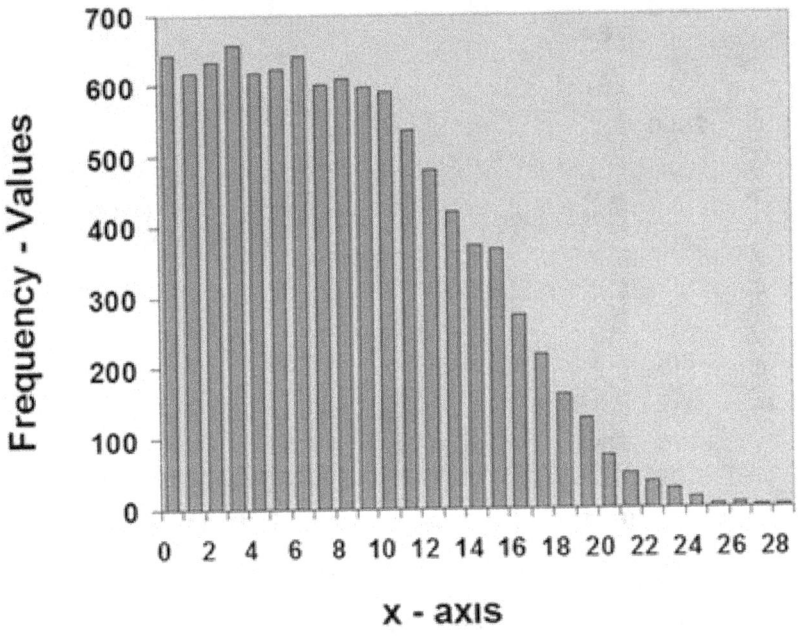

Figure 2.31: 10,000 Simulations of the Chain Uniform(0, Normal(17, 4))

that the small should be numerous and the big rare. Indeed, 10,000 computer simulations of this chain confirm the small is beautiful phenomenon as expected. Figure 2.31 depicts the histogram of 10,000 simulated values, clearly demonstrating that the small is overall more numerous than the big.

In another example, four Uniform Distributions are chained with regards to parameter **b** only, and with the ultimate **b** value fixed at 55, while parameter **a** is always fixed at 0. More concisely, this chain scheme is expressed as:

Uniform(0, Uniform(0, Uniform(0, Uniform(0, 55)))).

This long chain process can also be described as a step-by-step simulations scheme as follows: Simulate a single value from the Uniform(0, 55) and call it N; then simulate a single value from the Uniform(0, N) and call it M; then simulate a single value from the Uniform(0, M) and call it L; then simulate a single value from the Uniform(0, L) and call it K. This last K value is the final value of the

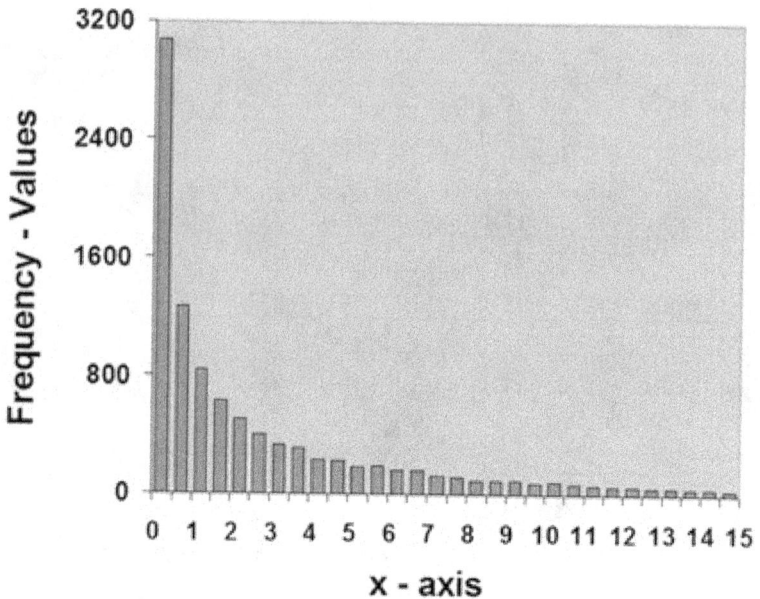

Figure 2.32:   10,000 Simulations of the Chain $U(0, U(0, U(0, U(0, 55))))$

entire chain in this single simulation. Such a detailed description of the process of the chain hints at how or why simulated values tend to crowd out towards 0 with each new step throughout the entire process. In other words, that $0 < K < L < M < N < 55$.

Figure 2.32 depicts the histogram of 10,000 simulated values of the chain of distributions Uniform(0, Uniform(0, Uniform(0, Uniform(0, 55)))). The small is clearly more numerous than the big here for this long chain, and more dramatically and consistently so as compared with the shorter chain of Uniform(0, Normal(17, 4)) in Figure 2.31.

The relevance of the chain of distributions model to real-life measurements and data sets becomes more apparent considering the causality in life and nature; the interconnectedness in the world; and the dependencies of some entities on other entities. All this leads to the conclusion that often some physical measurements serve as parameters for other physical measurements. For example, lengths and widths of rivers depend on average rainfall (being the parameter) and rainfall in turns depends on sunspots, prevailing winds, and geographical location, all serving as parameters of rainfall. Weights of people may depend on overall childhood nutrition, while nutrition in turns may depend on overall economic activity, which in turns depends on economic policy, war and peace, weather-related events such as droughts and flooding, and so forth. Exact cause and effect relationships in the deterministic realms such as in physics, chemistry, astronomy, biology, economics, and so forth, often lead to skewed results and scenarios which can be modeled on chains of distributions, and especially so when phenomena is aggregated and examined on a large scale. This is another chief cause (explanation) of why so often the small is more numerous than the big in the world.

Chapter 42

The Prevalence of the Lognormal Distribution Favoring the Small

An alternative explanation regarding the prevalence of the small is beautiful phenomenon in the physical sciences and real-life data sets refers to the fact that quite often such measurements closely mimic the Lognormal Distribution, and almost always with a high value of the shape parameter of over 1.0, implying high order of magnitude. The histogram of a data set with a high shape Lognormal configuration, falls off sharply on the right after a very brief and temporary rise on the left, hence overall, the small is numerous and the big is rare. For the minority of Lognormal Distributions with very low shape parameter values and the implied low order of magnitude such as those with values below 0.5 or 0.7 approximately, the small is beautiful phenomenon is absent, and the curve appears to mimic the Normal Distribution a great deal.

Indeed, the Lognormal Distribution is often the appropriate and fitting description of numerous physical, natural, and social phenomena. Some illustrative examples include the following:

> *Geology*: Several atmospheric chemical and physical properties, such as the distributions of the sizes of aerosols as well as clouds and related parameters of turbulent development.
> The concentration of elements and their radioactivity in the Earth's crust.

Medicine and Biology: Most animal and plant communities as well as the profusion of most species, namely the measures of size of living tissue such as weight, length of creatures, skin area, length of inert appendages of biological specimens that grow such as nails, teeth, hair, claws, horns, and so forth.

Pollution Affects: Distributions of sensitivity to fungicides pollution in any large populations.

Environment and Biology: Distribution of organisms and chemicals in the environment.

Disease: The time distribution of the appearance of lung cancer for cigarette smokers.

The abundance of bacteria on a variety of plant species.

Medicine: Number of side effects of any particular medicine.

Communicable Epidemics: The number of hospitalized cases as a measure of the spread.

Hydrology: Quantities of daily rainfall. Quantities of river water discharge volumes.

Chemistry: Sizes of particles, colloidal and polymer chemistry, and molar mass distributions.

Economics: Family and individual income distribution in populations.

The size of agricultural farms within any country.

Finance: Stock market indices worldwide, changes in exchange rate values, and price indices.

Social Sciences: The age of first marriage in countries of the Western civilization.

Linguistics: The number of letters per word and the number of words per sentence.

Human Behavior: Users' spent time on online articles.

Often, the Lognormal Distribution is obtained in repeated multiplications of random variables.

Hence, in a more profound sense, it is simply just one manifestation of the more generic tendency in multiplication processes leading to the small is beautiful phenomenon.

In addition, the Lognormal distribution is associated with natural growth processes driven by the accumulation of many small percentage changes and which can be interpreted as the random growth model regarding the last elements of multitude growth series mentioned in Chapter 37.

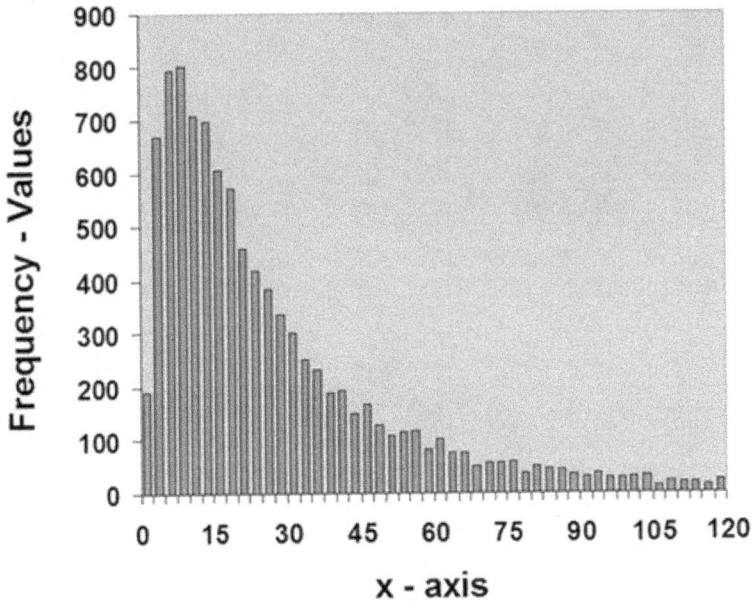

Figure 2.33:   Histogram of 10,000 Simulations of Lognormal Distribution

Figure 2.33 depicts the histogram of 10,000 simulated values from the Lognormal Distribution with location parameter 3 and shape parameter 1. In the context of relative quantities, the essential feature of the Lognormal Distribution is that it's asymmetrical and falling to the right overall; hence the small is more frequent than the big except for a very small portion of the data on the left. The brief rise in the beginning near 0 on the left over only a very narrow sub-interval is not significant in the grand scheme of things.

Chapter 43

The Prevalence of the Exponential Distribution Favoring the Small

The Exponential Distribution is often used to model the time between the occurrences of events in an interval of time, as well as the distances between events in space. The Exponential Distribution may be useful to model events such as:

- The number of minutes between eruptions for a certain geyser.
- The time it takes before your next telephone call.
- The number of minutes between customers who enter a certain shop.
- The time between customer calls to a help center.
- The time between goals scored in a football match.
- The time between successive failures of a machine.
- The distance between successive breaks in a pipeline.
- The time from diagnosis until death in patients with metastatic cancer.
- The time between meteors greater than one meter in diameter striking earth.
- The service times of agents, such as the time it takes for a bank teller to serve a customer.
- The time until a radioactive particle decays, or the time between clicks of a Geiger counter.
- The amount of time until the hardware of any computer fails.

- The amount of time you need to wait until the train arrives.
- Lifetime of electronic devices such as light bulbs or batteries, as they tend to fail randomly and independently over time.
- Interval time of emails or phone calls.
- Duration of remission of leukemia patients who are treated with a certain drug.

The Exponential Distribution is specified by a single (positive) parameter λ called Lambda.

Lambda is the event rate, namely the average number of events per unit time, or the average number of spacial occurrences per unit length.

Figure 2.34 depicts the histogram of 10,000 simulated values from the Exponential Distribution with Lambda parameter value of 666.6. In the context of relative quantities, the essential feature of the Exponential Distribution is that it is monotonically (consistently) falling to the right, hence the small is always more frequent than the big — and regardless of the value of parameter Lambda. The fact that

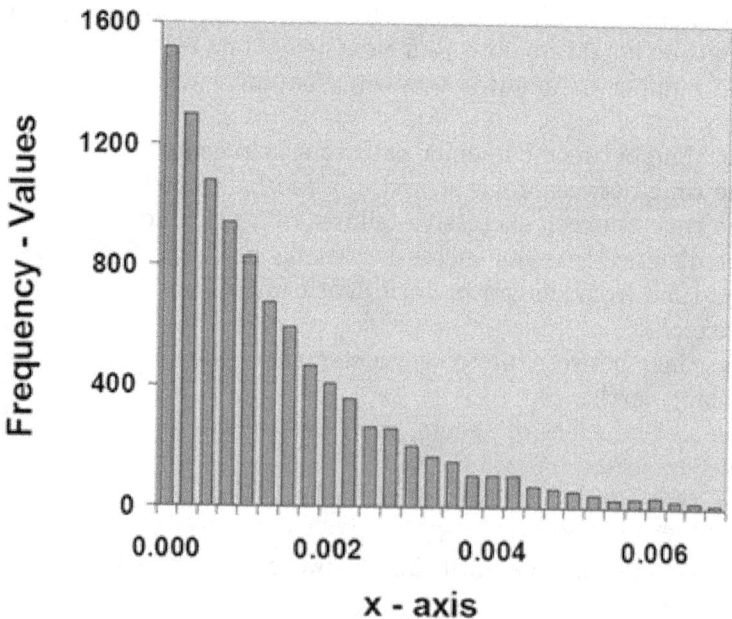

Figure 2.34: Histogram of 10,000 Simulations of Exponential Distribution

the Exponential Distribution falls monotonically and consistently to the right can also be seen in the expression for its Probability Density Function $\lambda e^{-\lambda x}$, or as in its quotient format $\lambda/e^{\lambda x}$ which indicates more clearly that the curve is inversely proportional to $e^{\lambda x}$ and thus inversely proportional to x (since Lambda λ is positive).

In the above simulations, the expression for the Cumulative Distribution Function is Probability$(X < x) = 1 - e^{-\lambda x}$. This leads to the relationship: RAND$() = 1 - e^{-\lambda x}$.

Solving for x we obtain: $x = -\text{LOGe}(1 - \text{RAND}())/(\lambda)$. Thus, 10,000 random values from the Uniform(0, 1) are generated via the function RAND() in MS-Excel, and each such value is inserted in the expression $-\text{LOGe}(1 - \text{RAND}())/(666.6)$ to arrive at these 10,000 Exponential values.

Random Real Partition is intimately connected with the Exponential Distribution. For example, an L-meter pipe is cut randomly into N parts via $(N - 1)$ random markings between 0 and L applying the Uniform Distribution. This implies the rate of $(N - 1)/(L)$ markings per unit distance, and pointing to the Exponential Distributions with $(N-1)/(L)$ Lambda parameter value.

In order to demonstrate the intimate relationship between Random Real Partition (as in Random Pipe Breaking) and the Exponential Distribution, a 15-meter pipe shall be cut randomly into 10,000 parts via 9,999 random markings between 0 and 15 applying the Uniform Distribution. This implies the rate of $(9,999)/(15)$ or 666.6 markings per meter. Therefore a comparison could be made between this random process and the Exponential Distributions with 666.6 Lambda parameter value discussed above. Indeed the reason the Exponential Distribution above was intentionally selected with 666.6 Lambda parameter value is to be able to make such a comparison.

In performing Random Real Partition here, 9,999 realizations are generated from the continuous Uniform(0, 15) via simulations using the expression $= 15 \times \text{RAND}()$, where RAND() is the function simulating values from the Uniform(0, 1). Values are ordered from low to high, 0 is added on the very left, and 15 is added on the very right. Then the difference data set is generated out of the above set, representing resultant pipe parts after the breakup. Certainly, this computer simulation process is absolutely distinct from the computer simulation process above of the Exponential Distribution

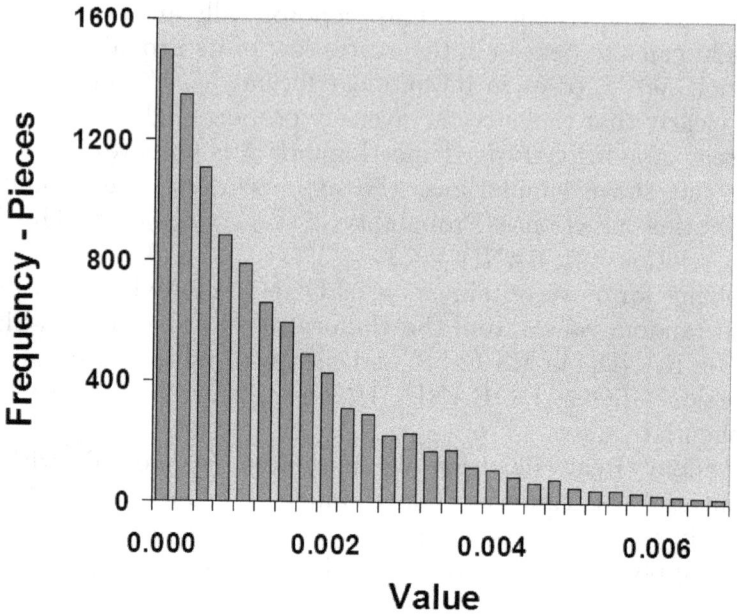

Figure 2.35: Histogram of 10,000 Parts of Random 15-Meter Pipe Breaking

with Lambda value of 666.6, yet both processes yield nearly identical results with only tiny and insignificant differences. This in a sense nearly corroborates the claim that they are both of the same nature, and that they express the same random process. Figure 2.35 depicts the histogram of these 10,000 simulated pipe parts.

Staring at these two histograms of Figures 2.34 and 2.35 gives the impression that they are almost of identical form, and nearly confirms the assertion that Random Real Partition and the Exponential Distribution are intimately related. Surely there are tiny random variations between the histograms, but such is the fate of all random processes. Mathematical Statistics provides a rigorous proof that Random Real Partition as in Random Pipe Breaking is simply the Exponential Distribution, with parameter λ Lambda being the average number of spacial occurrences per unit length (i.e. the number of random cuts per unit length).

In addition to the above mentioned real-life physical cases where the Exponential Distribution applies, another profound reason for the existence of the small is beautiful phenomenon in the natural world is the Exponential-like form of three essential and widely occurring distributions in Physics. These distributions are used extensively in Thermodynamics and Quantum Mechanics, having either the exact form of the Exponential Distribution or some very similar form of distribution.

The three distributions are: Boltzmann–Gibbs, Bose–Einstein, and Fermi–Dirac distributions.

Their distribution forms shall be compared with the Probability Density Function of the Exponential Distribution $\text{PDF}(X) = \lambda e^{-\lambda X}$ with $X > 0$, Lambda parameter $\lambda > 0$, and shown to be very similar if not identical.

The **Boltzmann–Gibbs distribution**, named after Ludwig Boltzmann and Josiah Willard Gibbs, describes the behavior of particles in the field of Thermodynamics according to classical physics and where the effects of Quantum Mechanics are ignored as insignificant. The distribution predicts the macro properties of matter such as pressure, volume, and temperature, via the statistical analysis of their micro (molecular or atomic) random behavior. The expression of the Boltzmann–Gibbs distribution is $F_{\text{BG}}(E) = \beta e^{-\beta E}$, with k as the Boltzmann constant, T as the temperature, $\beta = 1/kT$, while the independent variable E represents the energy of the system. The distribution here is exactly as the Exponential Distribution in form, and thus the histogram consistently falls to the right, manifesting the small is beautiful phenomenon.

The **Bose–Einstein distribution**, named after Albert Einstein and Satyendra Nath Bose, describes a quantum system of non-interacting bosons in the fields of Quantum Statistics and Thermodynamics. The expression of the Bose–Einstein distribution is $f_{\text{BE}}(E) \sim 1/(e^{\beta E} - 1)$, written with a proportionality sign instead of an equality sign. Although the form is slightly different from the form of the Exponential Distribution owing to the extra "-1" term in the denominator, yet the essential feature of a consistently falling histogram to the right is present here just as in the Exponential, especially when energy E is large rendering the "-1" term insignificant.

The **Fermi–Dirac distribution**, named after Enrico Fermi and Paul Dirac, describes the behavior of electrons involving the combined applications of the fields of Thermodynamics and Quantum Mechanics. The expression of the Fermi–Dirac distribution is $F_{\text{FD}}(E) = [\beta/\ln(2)] \times [1/(e^{\beta E} + 1)]$. Although the form is slightly different from the form of the Exponential Distribution owing to the extra "+1" term in the denominator, yet the essential feature of a consistently falling histogram to the right is present here just as in the Exponential, especially when energy E is large rendering the "+1" term insignificant.

The common feature across these three distributions above is that the algebraic expression is such that the distribution is inversely proportional to the variable E. The larger the value of E the higher is the value of the denominator and the lower is the value of the entire expression, yielding lower height on the histogram. This implies a tail in the histogram falling monotonically and consistently to the right, and thus the small is beautiful phenomenon. Consequently, this bestows a colossal scope for the small is beautiful phenomenon which pervades in the entire microworld of atoms, molecules, and elementary quantum particles throughout the universe.

Discovery of a Size Pattern: The General Law of Relative Quantities

Chapter 44

The Quest for Numerical Consistency in All Data Sets

In the first section of the book, the small is beautiful phenomenon was shown to manifest itself in a variety of measurements, data sets, game results, and physical entities. This was accomplished via the construction of a single histogram. Definitions of what exactly should constitute the "small" or the "big" were somewhat arbitrary, yet surely reasonable. The first bin on the left could be considered as "small", and the last bin on the right as "big", although admittedly all this depends on the arbitrary width of the bin. More significantly, adjustments and modifications of these size and bin definitions would certainly not lead to any revision of the overall conclusion that the small is indeed more numerous than the big, no matter how we construct the histogram. This is so whenever the histogram is generally falling, standing quite high on the left and hovering very low on the right in the grand scheme of things, in spite of the occasional mild upsets, tiny zigzags, and sporadic reversals.

Could an exact quantitative measure indicating by how much the small outnumbers the big be found such that the measure is roughly consistent across all data types obeying the small is beautiful principle? One such simplistic attempt at an exact measure shall be made in the next chapter. Unfortunately, the endeavor ends in a decisive failure as the measure varies greatly depending on data type; not showing any quantitative consistency across all cases and data sets.

Chapter 45

Division of Data Among Small, Medium, and Big Sizes

In this chapter, each data set under consideration shall be divided in its entirety into three size categories by constructing a histogram with only three bins designated as Small, Medium, and Big.

In order to divide a given data set into Small, Medium, and Big categories, it is necessary to decide on two border points, Border Point A and Border Point B, reasonably chosen within the range of the particular data under consideration. Consequently, all data points falling to the left of Border Point A are deemed as Small; all data points falling between Border Point A and Border Point B are deemed as Medium; and all data points falling to the right of Border Point B are deemed as Big. In other words, Border Points A and B are the threshold or cutoff points differentiating between the three sizes. This arrangement can be visualized in Figure 3.1 which depicts the classification of Small/Medium/Big via these two Border Points. The reason for the slightly lesser range allocated to the Medium shall be explained shortly. By convention, any data points falling exactly on Border Points A are deemed as Medium; and any data points falling exactly on Border Points B are deem as Big.

For smoothly spread data, it is only from the knowledge of the range of the data under consideration that we construct these two border points, and almost without utilizing any information regarding relative concentrations of values within the range provided by the data's density or histogram. For data with severely irregular spread, where the portions of the data on the margins are highly

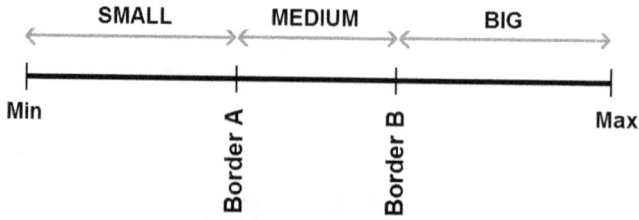

Figure 3.1:   Classification of Small, Medium, and Big via Two Border Points

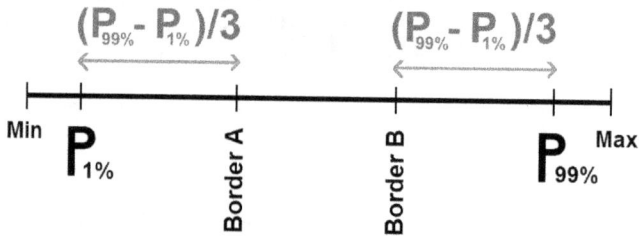

Figure 3.2:   The Creation of Two Border Points via the Exclusion of Top 1% and Bottom 1%

diluted so that the edges are considered outliers, these two border points are constructed in a way that attempts to ignore and omit these outliers from the calculations.

The most obvious way to choose Border Point A and Border Point B is by simply dividing the entire range, from its minimum to its maximum, equally into three sub-intervals, each having the same length, namely $(\text{Max} - \text{Min})/3$. Yet, in order to avoid excessive influence from outliers on the very margins of data, a more reasonable approach here is to start at the 1st percentile point and to end at the 99th percentile point, utilizing these percentile points as the two edges of the core 98% range of data. Figure 3.2 depicts the arrangement of the two Border Points A and B.

The 1% percentile point, or the first percentile point, denoted as $P_{1\%}$, is the value below which 1% of the ordered data may be found.

The 99% percentile point, or the 99th percentile point, denoted as $P_{99\%}$, is the value below which 99% of the ordered data may be found.

In what might seem as a paradox, once Border Point A and Border Point B are chosen with the deliberate exclusion of the outliers, the entire set of numbers in the data are now incorporated into the designation Small/Medium/Big, including even those outliers on the margins. In other words, after excluding the margins from the determination of Border Points A and B, the points falling between Min and $P_{1\%}$ are to be included at the end and are allocated to Small, and the points falling between $P_{99\%}$ and Max are to be included at the end and are allocated to Big. This seemingly contradictory attitude towards outliers is motivated by the realization of the potential for significant adverse effects on the Border Points A and B due to outliers; as opposed to the insignificant and very mild effects of adding only very few extra data points (outliers) to Small and Big. By including the outliers in the classification scheme at the end, we ensure that nothing within the entire data set is omitted or neglected. The diagrams of Figures 3.1 and 3.2 define the two border points as follows:

Border Point A $= P_{1\%} + (P_{99\%} - P_{1\%})/3$

Border Point B $= P_{99\%} - (P_{99\%} - P_{1\%})/3$

In one concrete numerical example, demonstrating the absolute necessity of avoiding the utilization of the maximum, minimum, and possible outliers in the construction of the two Border Points, a data set is imagined having 30,000 points which are approximately uniformly distributed on (5, 35), plus a single outlier value of 155. Allowing this outlier to influence decisions regarding Border Points A and B by partitioning the entire range from the minimum 5 to the maximum 155 into 3 equal sections of $(155 - 5)/3 = (150)/3 = 50$ width each, would cause the Small to be defined over (5, 55), and thus having the Small artificially earn nearly 100% of overall data portion. By using the 1st and the 99th percentiles, we arrive at a much more reasonable partitioning scheme along (5, 15.1) for the Small, [15.1, 24.9) for the Medium, and [24.9, 155) for the Big. It should be noted how Medium earns less range than either Big or Small.

For this data set, $P_{1\%} = 5 + 0.01 \times (35 - 5) = 5.3$, and $P_{99\%} = 5 + 0.99 \times (35 - 5) = 34.7$. It follows that Border Point A $= P_{1\%} + (P_{99\%} - P_{1\%})/3 = 5.3 + (34.7 - 5.3)/3 = 5.3 + 9.8 = 15.1$, and that Border Point B $= P_{99\%} - (P_{99\%} - P_{1\%})/3 = 34.7 - (34.7 - 5.3)/3 = 34.7 - 9.8 = 24.9$.

Medium is full of envy; it is bitterly complaining that its share on the entire range is less than that allocated to either Small or

Big; and that it is being discriminated against. Subsequently, the algorithmist points out to Medium that differences are really tiny; that its loss is no more than 2% of the overall range, and finally Medium reluctantly accepts the arrangement so as not to appear as obstructionist, still whispering to itself all sorts of old grievances and ridiculous accusations against the Small and especially against the Big, such as in the alleged maltreatments and discriminations of middle children in large families.

This algorithm shall be applied for most of the data sets, distributions, casino dice games, and other measurements and topics discussed in sections 1 and 2. The list of 24 such data sets or cases is outlined below, assigning a particular short name for each one in **boldface** characters.

Carbon Dioxide: CO_2 emissions by 216 sovereign states and territories in 2008, Chapter 16.

Rivers: List of 181 most significant rivers worldwide, longer than 1,000 kilometers, Chapter 7.

Exoplanets: Mass of the 1,404 known exoplanets as of September 21, 2016, Chapter 6.

Capitalization: Market capitalization values of 2,889 companies as of Oct 9, 2016, Chapter 12.

Population: USA 2009 population counts of 19,509 incorporated cities and towns, Chapter 9.

Wars: Death toll of 134 major wars from antiquity to the modern era, Chapter 10.

Molecules: Molar mass in a list of 2,175 chemical compounds, Chapter 3.

Pulsars: Spin frequency of 2,560 known pulsars as of December 2, 2016, Chapter 5.

Expenses: 987,492 expense bills of the State Of Oklahoma for the fiscal year 2011, Chapter 13.

Price List: The price list of the 14,914 items on sale for Canford Audio PLC, Chapter 14.

House Number: 23,633 addresses on Prince Edward Island in Canada, Chapter 15.

Primes: The set of 9,592 primes from 2 to 100,000, relating to Chapter 21; but this data set is not specifically discussed there.

Rock Breaking: Final set of random computer simulations with 8,192 pieces of 100-kilogram rock breaking in 13 stages, relating to Chapter 29; but this data set is not specifically discussed there.

Pipe Breaking: Final set of random computer simulations with 13,000 parts of 38-meter pipe breaking, relating to Chapter 30; but this data set is not specifically discussed there.

Four-Dice Products: All 1,296 possible four-dice multiplication events, relating to Chapter 22; but this data set is not specifically discussed there. There are $6 \times 6 \times 6 \times 6 = 1296$ possibilities of dice combinations; each is given equal probability in the game.

Exponential Growth: 5% exponential growth series from 7 to 101,643,854 in 338 periods, relating to Chapter 34; but this data set is not specifically discussed there.

30-Dice Dependency: 20,000 simulations of 30-dice selection games, relating to Chapter 40.

Uniform: 10,000 computer simulations from the Uniform(2, 5) Distribution.

Normal: 10,000 computer simulations from the Normal(17, 4) Distribution.

Short Chain: 10,000 simulations of Uniform(0, Normal(17, 4)), Figure 2.31, Chapter 41.

Long Chain: 10,000 simulations of U(0, U(0, U(0, U(0, 55)))), Figure 2.32, Chapter 41.

Lognormal: 10,000 simulations of Lognormal, location 3, shape 1, Figure 2.33, Chapter 42.

Exponential: 10,000 simulations of Exponential with Lambda 666.6, Figure 2.34, Chapter 43.

Earthquakes: The time in seconds between successive earthquakes worldwide for the entire year of 2012 in which 19,452 earthquakes occurred globally. See Kossovsky (2014) Chapter 11 for more details. This Earthquake data set which nicely obeys the small is beautiful principle has not been discussed in the previous sections.

The data sets of Primes, Rock Breaking, Pipe Breaking, Four-Dice Products, Exponential Growth, and 30-Dice Dependency are variations on the topics discussed in earlier sections. These new data sets differ slightly from the specific examples given in these sections by changing only the numerical parameters, limits, and/or the conditions of the underlying schemes and processes.

Figure 3.3 provides detailed results for these 24 data sets and cases. The column labeled "Data Size" refers to the total number of values within each data set or case. Percents of Small/Medium/Big are calculated with respect to Data Size, namely the percent of say Small occurrences within the entire data set. The last column labeled "B/S" refers to the division of Big by Small. The entire table is sorted from low to high according to the last column "B/S".

Symmetrical distributions and histograms are the antithesis of the small is beautiful phenomenon, hence the Normal and the Uniform in the bottom rows of Figure 3.3 are exceptional, and their size configurations stand in sharp contrast to the rest of the data sets and cases in the table above them. In addition, the case of Primes also stands apart and must be considered as a unique example since it refuses to fully obey the small is beautiful principle, and its distribution is derived from a particular regularity being inversely proportional to the logarithm of a given prime. Primes do not truly constitute a manifestation of the phenomenon as differences between sizes are very mild in comparison to the rest of the table, and in the limit, for much higher primes, the histogram is supposed to attain the flat and horizontal shape of the Uniform. Hence the focus of all the subsequent discussion and analysis would refer only to the first 21 data sets and cases — from Carbon Dioxide all the way to Short Chain — in the top 21 rows of the table.

The common feature across all these 21 data sets and cases is that they almost consistently reflect the small is beautiful phenomenon. In all of these 21 data sets and cases Small is more numerous than either Big or Medium, and there is not a single exception here to this strict and nearly universal rule! Also, in almost all of these 21 data sets and cases Medium is more numerous than Big, with the rare (significant) exceptions of Expenses and Capitalizations.

The great satisfaction in having successfully confirmed once again the small is beautiful phenomenon with these 21 data sets and cases by way of this new algorithm is fleeting, as it is immediately eclipsed

Data Set	Data Size	Small	Medium	Big	S%	M%	B%	B / S
Carbon Dioxide	216	211	2	3	98%	1%	1%	0.014
Expenses	967,682	942,885	10,959	13,838	97%	1%	1%	0.015
Rock Breaking	8,192	7,923	141	128	97%	2%	2%	0.016
Capitalization	2,889	2,796	43	50	97%	1%	2%	0.018
Population	19,509	18,662	488	359	96%	3%	2%	0.019
Exoplanets	1,404	1,327	50	27	95%	4%	2%	0.020
Price List	14,914	14,098	502	314	95%	3%	2%	0.022
House Number	23,633	22,172	805	656	94%	3%	3%	0.030
Lognormal	10,000	8,944	774	282	89%	8%	3%	0.032
Exponential Gr.	339	313	14	12	92%	4%	4%	0.038
Wars	134	125	4	5	93%	3%	4%	0.040
Long Chain	10,000	8,741	901	358	87%	9%	4%	0.041
4-Dice Products	1,296	1,102	146	48	85%	11%	4%	0.044
Pulsars	2,560	2,293	160	107	90%	6%	4%	0.047
Earthquakes	19,451	15,764	2,830	857	81%	15%	4%	0.054
Pipe Breaking	12,999	10,258	2,147	594	79%	17%	5%	0.058
Rivers	181	151	21	9	83%	12%	5%	0.060
Exponential	10,000	7,763	1,748	489	78%	17%	5%	0.063
Molecules	2,175	1,479	567	129	68%	26%	6%	0.087
30-Dice Depend.	20,000	12,902	5,043	2,055	65%	25%	10%	0.159
Short Chain	10,000	4,685	3,799	1,516	47%	38%	15%	0.324
Primes	9,592	3,565	3,037	2,990	37%	32%	31%	0.839
Normal	10,000	2,294	5,677	2,029	23%	57%	20%	0.884
Uniform	10,000	3,311	3,249	3,440	33%	32%	34%	1.039

Figure 3.3: Resultant Size Configurations for Small, Medium, and Big

by the profound disappointment of not being able to obtain any consists and exact measure (numerical pattern) of the phenomenon across all or even most of these data sets and cases. Unfortunately, the algorithm yields too much variation in its measure.

Percent of Small varies widely, fluctuating between 47% and 98%.
Percent of Big varies widely, fluctuating between 1% and 15%.
The ratio Big/Small varies widely, fluctuating between 0.014 and 0.324.

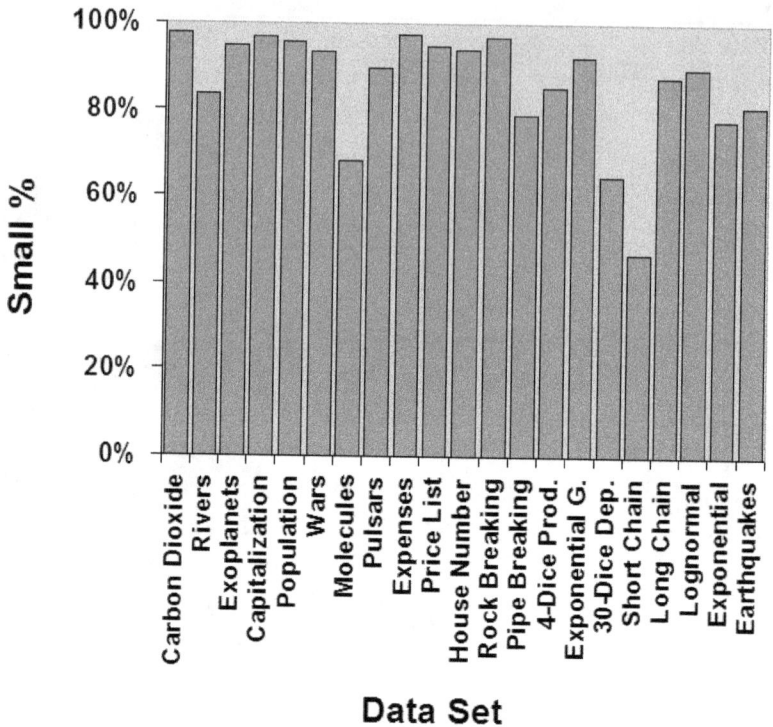

Figure 3.4:   Percents of Small for the 21 Data Sets and Cases

Figures 3.4 and 3.5 depict the percents of Small and the percents of Big for the 21 data sets and cases in the application of the algorithm, using the original "order" above of their descriptions without any sorting. The chart of Big/Small (which is not shown here) appears almost identical to Figure 3.5 of percents of Big, and as such it does not aid visually in any way, nor does it provide any new insight.

30-Dice Dependency and Short Chain deviate somewhat from the behavior of their data peers, not participating whole-heartedly in the small is beautiful trend, having lower percents for Small and higher percents for Big as compared with the rest of the group. Ignoring 30-Dice Dependency and Short Chain, the fluctuation diminishes a bit for the shorter list of 19 data sets and cases, as Small varies only from 68% to 98%, while Big varies only from 1% to 6%.

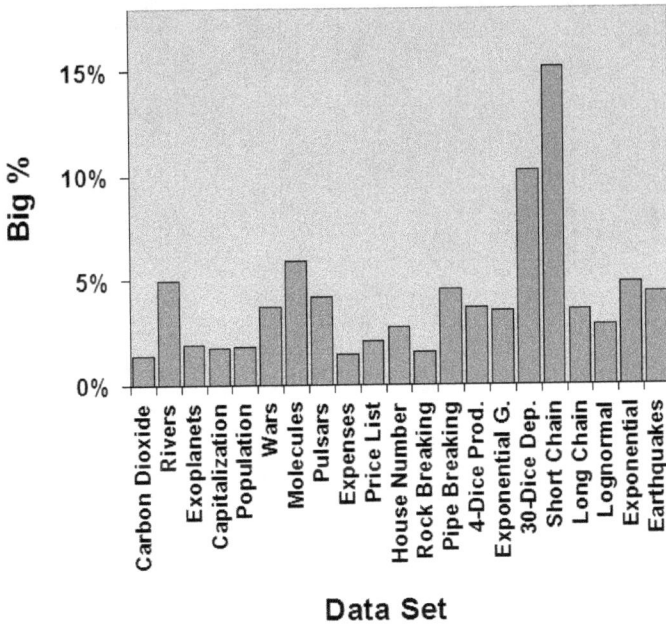

Figure 3.5: Percents of Big for the 21 Data Sets and Cases

It should be noted that 30-Dice Dependency scheme is readily modeled on the short chain Uniform(1, Uniform(2, 31)) which is a bit similar to the Short Chain Uniform(0, Normal(17, 4)), hence in some limited sense these two schemes actually constitute almost a singular case.

Chapter 46

Constructing a Set of Infinitely Expanding Histograms

The decisive failure to obtain a consistent numerical measure across these 21 data sets and cases calls for a radical change in our approach regarding how relative quantities and sizes should be measure, so that a universal or nearly universal numerical result could be obtained.

There are two intrinsic features in the above three-bin algorithm that need to be abolished. The first feature is having a shifting histogram, constructed to fit the particular data set on hand by aiming at the particular range of the data. Clearly, each three-bin histogram spans different parts of the horizontal x-axis depending on the data set under consideration. The second feature is lazily constructing only one global histogram, attempting to calculate relative quantities and sizes in one fell swoop, instead of laboring hard and long by constructing a large set of repeated local histograms which could then be aggregated as one singular result. The motivation for such construction utilizing numerous histograms is to keep checking local relative quantities within all sub-ranges of the data, taking their distinct quantitative pulses everywhere along the x-axis, followed by the simple aggregation of all these pulses into one decisive result. Hence, instead of constructing a singular histogram with three bins of distinct widths and distinct x-axis focus, constructed subjectively and differently for each particular data set depending on the values of the minimum, maximum, first percentile point, and 99th percentile point, and so forth, what is necessary here is an objective, autonomous, fixed, and universal set of numerous histograms, to be positioned onto the x-axis from zero onward to infinity, without regard to any particular data set [and free from risks due to any possible outliers]. A common set of numerous histograms would be constructed for all data sets and for all cases, with the hope of achieving a measure of universality and consistency. For some deep-rooted mathematical reasons, the width of these histograms should be constantly expanding with each new construction, beginning with very narrow width near the 0 origin on the left side, and becoming ever wider on the right side of the x-axis towards infinity.

Figure 3.6 depicts the sketch of one such construction of repeated three-bin histograms, showing only the first three histograms in the whole scheme; neglecting to show the rest of the numerous and much wider histograms on the right, for brevity and lack of space.

In the scheme of Figure 3.6, each histogram comes with a fixed number of three equal bins called A, B, and C, all having the same width. The sizes of the histograms are expanding (doubling) with each cycle, so that the bin width is different for each histogram; some are very short and narrow (on the left), while others are very long and wide (on the right). The overall goal is to measure proportions of data falling within each of the three bins, A, B, and C, via the aggregation of all these histograms. In other words, once histogram construction is completed, we add all the results of all the bins designated as A as one grand sum; we add all the results of all the bins designated as B as one grand sum; and then we add all the results of all the bins designated as C as one grand sum. These three grand sums of A, B, and C, are the final set of three numbers determining relative frequency for the three sizes.

Naturally, bin A is thought of as Small, bin B as Medium, and bin C as Big, even though to the right of each bin C there exists

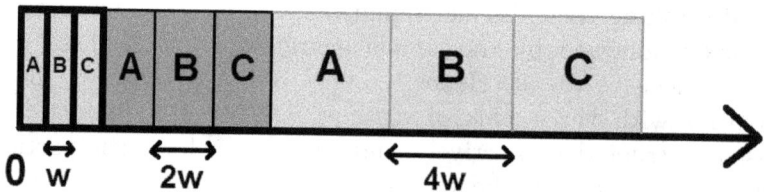

Figure 3.6:   Three-Bin Histograms with a Factor of Expansion of 2

a neighboring bin A (of subsequent histogram) pertaining to bigger values. Yet, such a vista is justified because within each histogram having the three bins A, B, and C, values within bin A are smaller than values within bin B or bin C, and values within bin C are bigger than values within bin B or bin A. On a more profound level, there exists a strong positive correlation between a skewed histogram of the data falling to the right and results in the expanding bin scheme favoring A over B as well as favoring B over C. If the histogram of the data set under consideration is overall flat, or if it is rising in the aggregate, then grand sum for all bins A is not larger than grand sum for all bins B, and grand sum for all bins B is not larger than grand sum for all bins C. Yet, if the data histogram is falling in the aggregate, then grand sum for all bins A is indeed larger than grand sum for all bins B, and grand sum for all bins B is indeed larger than grand sum for all bins C. Therefore, the relative proportions of these grand sums of bin A, B, and C tell us indirectly about the overall shape of the histogram as well as about relative quantities and relative sizes for the data set under consideration in its entirety.

In general, the scheme is of repeated cycles of local histograms with **D number of bins** each, and where bin width is expanding at **an inflation factor F**, followed by the aggregation of all of the histograms into a singular set of proportions. Hence, in Figure 3.6, the arrangement is that of a three-bin scheme with an inflation factor of 2, so these two parameters are D = 3 and F = 2. The entire system starts from the 0 origin with an initial infinitesimally small bin size W. In the second cycle, the width of each bin is 2W. In the third cycle the width of each bin is 4W. In the fourth cycle the width of each bin is 8W, and so forth.

Why is it necessary to expand the width of the bins with each consecutive histogram?

Why cannot a system with identical histograms having fixed bin width yield the desired results?

The heuristic argument here is that perhaps the phenomenon can be better understood if one acknowledges that a great deal of real-life data relates to or springs from multiplication processes, so that numbers are "rapidly expanded" and are "stretched out" along the x-axis in a multiplicative manner much as was seen in Figure 2.17 and in Figure 2.19 regarding the accelerated march of exponential growth series. If other (non-multiplicative) types of typical real-life data sets also resemble such accelerated quantitative configuration, then a (slow) system with identical histograms having fixed bin width simply cannot "keep up" with such rapid pace of the outlay of the data itself and thus could not serve as an appropriate measuring arrangement. If the *measured* data accelerates quantitatively, then the *measuring* bin scheme should also accelerate accordingly.

Why is it necessary to start the first cycle with such tiny bin width of infinitesimal dimension? Surely, if a very crude histogram with thick bins is applied in the first cycle, then most likely results would vary according to the subjective value of the width chosen relative to the objective values of existing data sets. Instead, the scheme is declared to be more universal and consistent via the objective choice of applying initially an infinitesimally small width — relative to all existing, or potentially existing data sets. Clearly, an extremely large and crude choice of W selected deliberately relative to a given data set could yield a misguided scheme that places all or most of the data points within the first bin in the first cycle, therefore artificially enabling the small to obtain a decisive albeit undeserving victory over all other sizes. Such subjective and arbitrary scheme does not inform us about the true and objective status of the data configuration and therefore should be avoided.

How small should the first width W be made in practice, so that statisticians and data analysts could actually implement a computer program to measure real-life data sets without worrying about highly abstract concepts such as the infinitesimal, the universal, or the objective? A straightforward and practical rule of thumb is to choose the value of W as say one-millionth (0.000001, namely 10^{-6}), assuming this small value is less than one-tenth or one-hundredth of the minimum of all the minimums in the collection of the data sets to be examined. If one millionth is not much less than the minimum of all the minimums, than an even smaller value should be chosen accordingly, say one-trillionth (0.000000000001, namely 10^{-12}). In addition, the scheme should be made to automatically stop whenever the maximum of all the maximums in the collection of the data sets to be examined is encountered by the computer, since there would be no more values to incorporate beyond this point.

Since it should always be assumed here that we are not dealing with negative numbers and that all the data sets are of positive numbers exclusively, the start from the origin 0 ensures that all corners and locations on the x-axis are covered; that all potential quantities are incorporated into the scheme and are accounted for.

The entire arrangement of (abstractly) infinite or (practically) numerous histograms, starting from the 0 origin, and with expanding bin widths, shall be coined **"bin scheme"**.

Chapter 47

Numerical Consistency in Bin Schemes for 15 Real-Life Data Sets

Let us consider the application of the bin scheme to some of the 24 data sets and cases outlined in Chapter 45.

Within the original 24 data sets and cases, **Normal**, **Uniform**, and **Primes** should definitely be excluded, since they adamantly refuse to obey the small is beautiful principle.

In addition, it might be prudent to exclude **30-Dice Dependency** and **Short Chain** from the bin schemes. The fact that both refused to fully participate in the small is beautiful trend measured as percentages as shown earlier in Figures 3.4 and 3.5 should not be any reason to exclude them from the different quantitative measure of the bin scheme, yet, since it typically takes at least 3 to 4 sequences for chains of distributions to (almost fully) converge to the small is beautiful phenomenon, these two very short chains of only two sequences should perhaps be excluded.

Since **Four-Dice Products** represents half-baked or incomplete multiplicative process, where resultant order of magnitude is actually not quite high enough in the context of the small is beautiful phenomenon, it might be wise to exclude this data set as well. Order of magnitude plays a crucial rule in facilitating the drive towards the small is beautiful phenomenon in multiplication processes, and order of magnitude is a bit lacking here for this Four-Dice Products. This four-dice game could have been considered as a more appropriate multiplicative process in this context had the game involved numerous six-sided dice instead of merely four dice, say 10 dice or 15 dice, or if dice with many more sides were involved, such as say 36-sided or 45-sided dice, and so forth — all of which yield greater resultant order of magnitude and stronger convergence to the small is beautiful phenomenon.

Moreover, **Rivers**, **Wars**, and **Carbon Dioxide**, with only 181, 134, and 216 values respectively, should also be excluded perhaps for lack of sufficient number of data points. Small data sets can be easily swayed by exogenous random factors totally unrelated to the small is beautiful trend.

Hence in summary, out of these 24 original data sets and cases, 9 should be excluded; leaving for the analysis only 15 data sets and cases which fully and wholeheartedly embrace the small is beautiful principle.

The above exclusions should not be thought of as false justifications in order to eliminate embarrassing results which could show significant deviations from the main trend here. There is solid mathematical reasoning which clearly supports the exclusion of these nine data sets and cases. As mentioned in Chapter 25, no claim is made here that all things in the universe are quantitatively configured as such that the small is favored. There are some counter examples and exceptions to the phenomenon. More significantly, there are many cases and processes which strictly adhere to the small is beautiful principle, yet lack an exact and consistent quantitative pattern. Therefore, by excluding those data sets with a fluctuating degree of loyalty to the small, a more exact and steady quantitative measure is obtained for those data sets with a steady degree of loyalty to the small. At first, all 24 data sets and cases shall be calculated according to the bin schemes and displayed in the tables clearly, without any exclusion a priori. Then, diagrams and conclusions shall refer to and include only those 15 data sets and cases having that consistent quantitative pattern.

The first bin scheme A is made with 5 bins for each histogram, constantly expanding by 2, namely doubling with each new cycle, having an initial small bin width of 0.00000000003, and automatically stopping when bin width reaches 40,000,000,000,000. The motivation for such extraordinarily small bin width is to establish a scheme compatible with all of these 24 data sets and cases by starting from a value well below the minimum of all the minimums. The stop at 40 trillion — which easily exceeds here the maximum of all the maximums — guarantees that all data points are covered by the scheme. There is no need to continue beyond 40 trillion where nothing will ever be found. The second bin scheme B is made with three bins for each histogram, constantly expanding by 3, having an initial small bin width of 0.00000000003, and automatically stopping at 40,000,000,000,000.

Bin Scheme A:

$D = 5$, $F = 2$, $W = 0.00000000003$, stop at 40,000,000,000,000.

Bin Scheme B:

$D = 3$, $F = 3$, $W = 0.00000000003$, stop at 40,000,000,000,000.

Figure 3.7 depicts the numerical results of Bin Scheme A for all the 24 data sets and cases, and partitioned into two classes according to consistency of results. The upper class of 15 data sets and cases is with superb numerical consistency, steadily displaying almost the same measure of quantitative configuration. The lower class of nine data sets and cases displays widely varying and inconsistent measure of quantitative configuration. Figure 3.8 depicts the numerical results of Bin Scheme B for all the 24 data sets and cases in the same fashion.

Remarkably, results [for the 15 upper class data sets] are almost the same for all practical purposes when other values of initial bin width W are used, so long as W is quite small. Formally in mathematical terms this condition is called "W-invariance", meaning that results do not vary under transformations or changes in the value of W. Hence, the results in Figures 3.7 and 3.8, for example, are [for all practical purposes] independent of W. Only when crude and too large W value is used that results could be severely distorted due to "capture" of large proportion of data by the first cycle, and in particularly by the first bin of the first cycle perhaps. In conclusion: only the combination of the pair of D and F values determines bin

Data Set	Bin A	Bin B	Bin C	Bin D	Bin E
Exoplanets	26.2%	22.6%	18.8%	17.1%	15.2%
Capitalization	26.4%	21.6%	19.5%	16.9%	15.5%
Population	26.2%	22.3%	19.3%	17.6%	14.7%
Molecules	26.9%	20.9%	19.4%	17.0%	15.8%
Pulsars	25.4%	23.2%	18.3%	17.6%	15.4%
Expenses	24.5%	23.5%	19.6%	15.4%	17.0%
Price List	25.7%	22.5%	19.7%	16.1%	16.0%
House Number	26.3%	21.7%	19.7%	16.5%	15.8%
Rock Breaking	26.2%	22.8%	18.5%	17.3%	15.2%
Pipe Breaking	26.4%	21.8%	19.0%	17.2%	15.6%
Exponential Gr.	26.5%	21.8%	19.5%	17.1%	15.0%
Long Chain	26.4%	22.1%	19.3%	16.7%	15.5%
Lognormal	26.5%	21.7%	19.2%	17.4%	15.3%
Exponential	26.5%	21.4%	19.3%	17.5%	15.4%
Earthquakes	26.8%	22.2%	19.2%	16.7%	15.2%
Wars	34.3%	19.4%	17.2%	13.4%	15.7%
Rivers	25.4%	16.0%	13.8%	27.6%	17.1%
Carbon Dioxide	29.2%	19.4%	22.2%	14.8%	14.4%
Short Chain	25.7%	22.5%	20.6%	16.9%	14.3%
30-Dice Depend.	24.6%	13.3%	40.4%	8.7%	13.1%
4-Dice Products	28.4%	13.4%	23.8%	19.1%	15.2%
Primes	31.9%	17.4%	17.1%	16.9%	16.7%
Uniform	16.9%	16.7%	19.5%	26.1%	20.8%
Normal	23.2%	16.4%	20.0%	21.4%	18.9%

Figure 3.7:   Results for 24 Data Sets and Cases $-D = 5$, $F = 2$

results, and nothing else matters, so long as very small and refine W value is used. Formally, an initial infinitesimally small bin size W must be used.

In order to clearly demonstrate this generic feature in all bin schemes, US Population data set shall be repeatedly examined with bin scheme $D = 5$ and $F = 2$ using a variety of W values:

Data Set	Bin A	Bin B	Bin C
Exoplanets	48.4%	31.1%	20.4%
Capitalization	46.3%	29.5%	24.2%
Population	45.3%	31.4%	23.3%
Molecules	46.6%	30.9%	22.4%
Pulsars	45.1%	31.1%	23.8%
Expenses	46.1%	30.6%	23.4%
Price List	47.2%	31.1%	21.6%
House Number	48.6%	28.6%	22.8%
Rock Breaking	47.4%	30.0%	22.7%
Pipe Breaking	46.5%	30.1%	23.4%
Exponential Gr.	46.6%	31.3%	22.1%
Long Chain	46.8%	30.1%	23.2%
Lognormal	47.3%	30.7%	22.1%
Exponential	46.2%	30.9%	23.0%
Earthquakes	46.7%	30.2%	23.1%
Wars	52.2%	27.6%	20.1%
Rivers	64.6%	20.4%	14.9%
Carbon Dioxide	43.1%	37.0%	19.9%
Short Chain	45.5%	28.9%	25.6%
30-Dice Depend.	46.1%	36.4%	17.5%
4-Dice Products	46.5%	26.1%	27.5%
Primes	44.5%	28.1%	27.4%
Uniform	37.9%	30.8%	31.4%
Normal	71.5%	17.8%	10.8%

Figure 3.8: Results for 24 Data Sets and Cases $-D = 3$, $F = 3$

$W = 0.00492 \quad \{26.2\%, 22.6\%, 19.0\%, 17.0\%, 15.2\%\}$

$W = 0.00516 \quad \{26.3\%, 22.6\%, 19.2\%, 16.6\%, 15.4\%\}$

$W = 0.13900 \quad \{26.3\%, 22.1\%, 19.6\%, 16.7\%, 15.3\%\}$

$W = 0.20500 \quad \{26.4\%, 22.1\%, 19.4\%, 16.9\%, 15.2\%\}$

$W = 0.70 \quad \{26.3\%, 22.5\%, 19.2\%, 17.0\%, 15.0\%\}$

$W = 1 \quad \{26.0\%, 22.4\%, 19.3\%, 17.6\%, 14.7\%\}$

$W = 500 \quad \{42.8\%, 24.3\%, 14.5\%, 10.3\%, 8.2\%\}$

Clearly, results for the top 6 lines vary only very little for these different values of W, and are basically pointing to the same set of proportions. The smallest city in the US is with only one person as his or her unique official resident, constituting the absolute possible minimum size here, thus setting $W = 1$ could not be considered as too large, and it still yields almost the correct result. On the other hand, when a crude bin scheme with $W = 500$ is applied, results are totally and dramatically distorted as seen in the last line above.

Figures 3.9 and 3.10 visually depict the numerical results of the two bin schemes, but only for those 15 data sets and cases with consistent quantitative results. Each data set or case is depicted as a singular histogram expressing the final result of the bin scheme, with the first left-most bin (colored black) representing Bin A, the second left-most bin (colored grey) representing Bin B, and so forth. Each successive bin, from A, to B, to C, and so forth, alternates color from black, to grey, to black, and back to grey, and so forth. Figures 3.9 and 3.10 are derived directly from the numerical information in the upper class of Figure 3.7 and the upper class of Figure 3.8. The

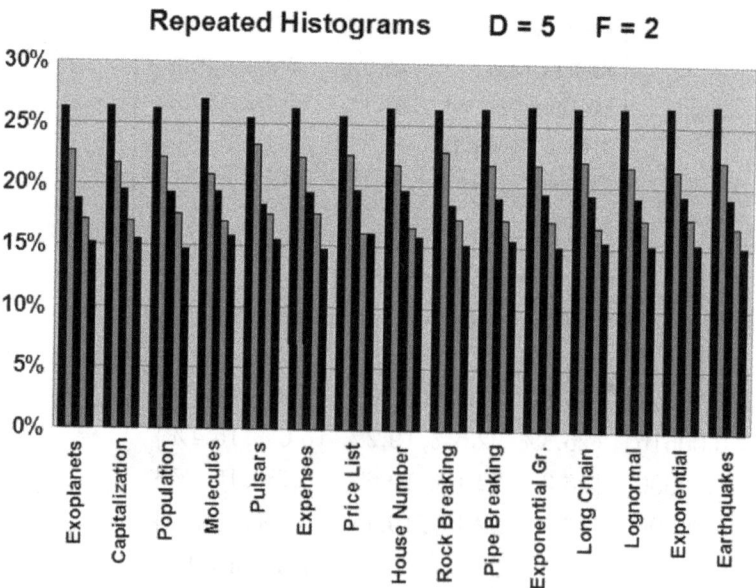

Figure 3.9: Pattern Found in the Consistent Results of 15 Data Sets — Bin Scheme $D = 5$, $F = 2$

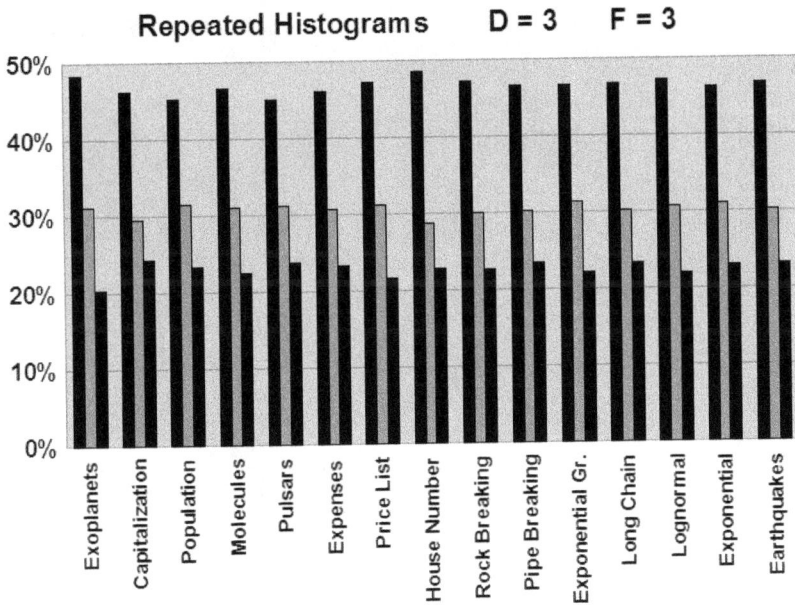

Figure 3.10: Pattern Found in the Consistent Results of 15 Data Sets — Bin Scheme $D = 3$, $F = 3$

role of Figures 3.9 and 3.10 is to provide visualization in the search for evidence of consistency in bin scheme results across a variety of data sets.

The bin scheme had succeeded superbly! A remarkable consistency has been found!

Several other bin schemes with different D and F values performed on these 15 data sets and cases also yielded highly consistent results.

The fact that the histogram of the bin scheme result for each data set or case (in the list of 15 items) is positively skewed with a tail falling to the right, is another strong confirmation of the small is beautiful phenomenon, since it could only been gotten when the histogram of each data set or case itself is also positively skewed in the aggregate, with a tail falling to the right for the vast majority of the range.

Chapter 48

The Quest for the General Mathematical Expression for All $D\&F$ Cases

Now that a clear trend of relative quantities has been found, with nearly exact numerical results for all 15 data sets and cases, the next challenging quest is to find the mathematical expression of a general law that would predict proportions of bin schemes for any given combination of D and F values. Further encouragement supporting and motivating this quest is found in the fact that this nearly exact trend is also confirmed in numerous other real-life physical and scientific data sets, far beyond the confines of the above 15 data sets and cases, constituting perhaps the only interdisciplinary phenomenon out there uniting diverse fields such as physics, thermodynamics, sub-atomic physics, astronomy, cosmology, chemistry, geology, biology, economics, demographics, and other disciplines. Surprisingly, this consistent quantitative pattern is found almost everywhere!

The goal of the scientist or the mathematician is to avoid the need to state numerous laws for all types of special cases; a law for the particular case of $D = 5$ and $F = 2$ for example; and another law for the particular case of $D = 3$ and $F = 3$; and to performed the endless and tedious task of stating numerous laws over and over again for all sorts of D and F combinations (perhaps by taking the average of the 15 results in each scheme). Rather, the quest is to come up with a generic mathematical expression that would encompass all possible

D and F cases in one fell swoop. In other words, the goal is to be able to predict bin proportions with high level of accuracy by simply knowing the values of D and F in the scheme and inserting these two values in a singular ready-to-use mathematical expression.

One possible approach in attempting to achieve this goal is to empirically test our 15 data sets and cases with a large collection of bin schemes for a wide variety of D and F values, creating a long table of D and F values along with the associated proportional results (the average of the 15 sets), and then attempting to find out **inductively** what is the best or the most fitting mathematical expression. Not much accuracy is lost if we choose just one large data set among them as a good representative of all 15 sets, in order to ease and speed up calculations.

As one concrete example of the inductive approach, a variety of bin schemes are explored for the Earthquakes data set, using several combinations of D and F values as shown in Figure 3.11. Since the minimum value in the Earthquakes data set is 0.01, the value of 0.002

D	F	Bin A	Bin B	Bin C	Bin D	Bin E	Bin F	Bin G
3	2	41.3%	32.1%	26.7%				
3	3	46.1%	31.0%	22.9%				
3	4	50.0%	29.9%	20.1%				
3	5	53.1%	27.8%	19.1%				
3	6	55.9%	26.4%	17.7%				
3	11	63.6%	22.1%	14.3%				
4	2	32.1%	26.7%	22.2%	19.0%			
4	3	37.2%	26.5%	19.9%	16.5%			
4	4	41.4%	25.9%	18.4%	14.3%			
4	5	43.3%	24.6%	18.0%	14.2%			
4	12	53.5%	21.0%	14.9%	10.7%			
5	2	26.7%	22.2%	19.0%	16.9%	15.2%		
5	5	36.5%	22.5%	17.0%	13.3%	10.6%		
5	9	42.0%	22.5%	15.6%	11.1%	8.8%		
7	2	19.0%	16.8%	15.2%	13.6%	13.1%	11.5%	10.7%
7	4	25.9%	18.4%	14.3%	13.0%	10.6%	9.8%	8.0%
7	8	32.6%	19.0%	13.5%	11.1%	9.3%	8.0%	6.4%

Figure 3.11: A Variety of Bin Schemes with Selected D and F Values — Earthquakes Data

for the initial bin width W is used in all of these bin schemes, so as to make sure all the schemes start well below this minimum value.

As expected, lower rank bins to the left are taller than higher rank bins to the right, so that the proportion in Bin A is always more than the proportion in Bin B, and the proportion in Bin B is always more than the proportion in Bin C, and so forth. Proportions are monotonically decreasing as higher rank bins of bigger values to the right are considered.

Clearly, for any fixed D value, larger F values yield skewer results where the small appears even more numerous. In other words, for any fixed D value, F and skewness are positively correlated. For example, for a 3-bin scheme in the top panel of Figure 3.11, $F = 2$ endows bin A the proportion 41.3%, $F = 4$ endows bin A the proportion 50.0%, while $F = 11$ endows bin A the proportion 63.6%. The converse could be stated for bin C, which earns progressively less for each larger value of F. That higher F values induce more dramatic effects of the small is beautiful phenomenon for any fixed D value is a general principle, true for all types of data with consistent bin results.

Chapter 49

The Postulate on Relative Quantities

Philosophically, a superior approach would not attempt to find out *inductively* what is the best or most fitting mathematical expression, but rather argue this by way of a conceptual postulate which would then lead to the mathematical expression *deductively* — all the while closely agreeing with empirical results from real-life physical data sets. In addition, since the inductive method here appears difficult to implement, the deductive approach represents the only realistic hope of discovering any mathematical law capable of fitting our empirical results.

The Postulate: The generic pattern in how relative quantities are found in nature is such that the frequency of quantitative occurrences is inversely proportional to quantity.

Inverse proportionality means that variable Y decreases as X increases, and that $Y = K/X$, for some constant number called K. The postulate implies that the histogram is falling to the right at a very particular rate, and that the height is constantly being reduced by half whenever quantity X is doubled. In other word, if quantity Q is twice as big as quantity P, namely $Q = P + P = 2P$, then the frequency of occurrences of Q is $K/Q = K/(2P) = (1/2)(K/P)$, and which is much rarer, being only 50% of the frequency of P calculated as K/P. Formally the postulate implies that:

Height of Histogram $=$ Constant/Quantity.

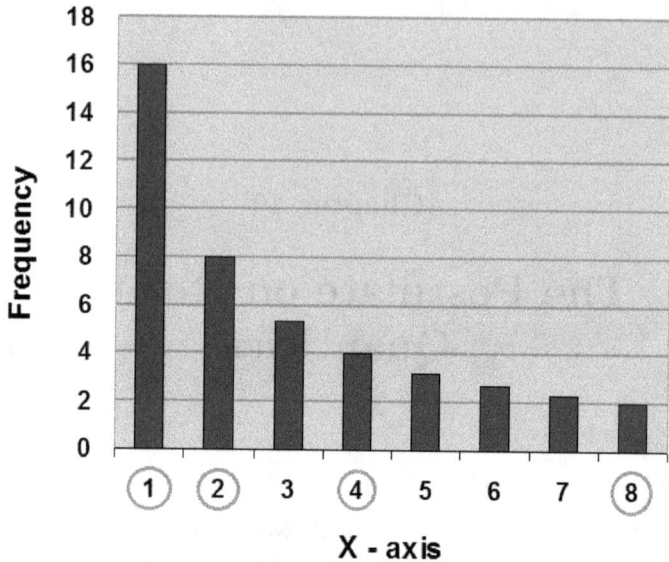

Figure 3.12: Inverse Proportionality as in $16/X$ — the Doubling of X Reduces Height by Half

Figure 3.12 depicts one histogram in the spirit of the postulate. Indeed, the height at each point is simply $16/X$. As we move from 1 to 2, doubling the quantity, height is reduced by a half, from 16 to 8. As we move from 2 to 4, doubling the quantity, height is reduced by a half, from 8 to 4. As we move from 4 to 8, doubling the quantity, height is reduced by a half, from 4 to 2.

An alternative postulate would assume that height is inversely proportional to X squared, or to X cubed, such as in Height = Constant/Quantity^2 or as in Height = Constant/Quantity^3, but these assumptions imply some very dramatic and rapid fall in the histogram, contrary to the general empirical evidences provided by the above 15 data sets and cases which hint at much milder and slower fall in the histogram, roughly close to the rate of Height = Constant/Quantity in the aggregate. Yet, a closer scrutiny of the rapidity in the fall in the histograms of the above data sets and cases shows local fluctuations and a great deal of variability, as 14 out of these 15 sets [i.e. all except Exponential Growth] are with a consistent gradual development in the shape of their histogram, from very low values where the histogram is usually rising very briefly, followed by an approximate flat and uniform histogram for a very brief range after that (and well before the average value is encountered), and finally to the consistent and long fall in the histogram to the right for most of the range. Attempting to aggregate and average out the very brief rise on the left and the very long fall on the right points approximately to an overall fall as in the rate of Constant/Quantity — in accordance with the postulate. The one data set standing apart from all the rest is Exponential Growth, which shows a near perfect agreement with the postulate, as its histogram falls off steadily at almost exactly the same rate as Constant/Quantity, without any temporary ups and downs, all the way from its lowest values to its highest values, lending the postulate a measure of credibility, and injecting confidence in its eventual success.

An elegant feature associated with the postulate is that total quantity is the same for all sizes.

Total quantity for a given size is simply the number of existing pieces or items of that size multiplied by the size itself. In other words, total quantity per size is the total value generated by summing up all the quantities having that particular size. This particular feature can be seen clearly in Figure 3.13 where total quantity for size 8 is

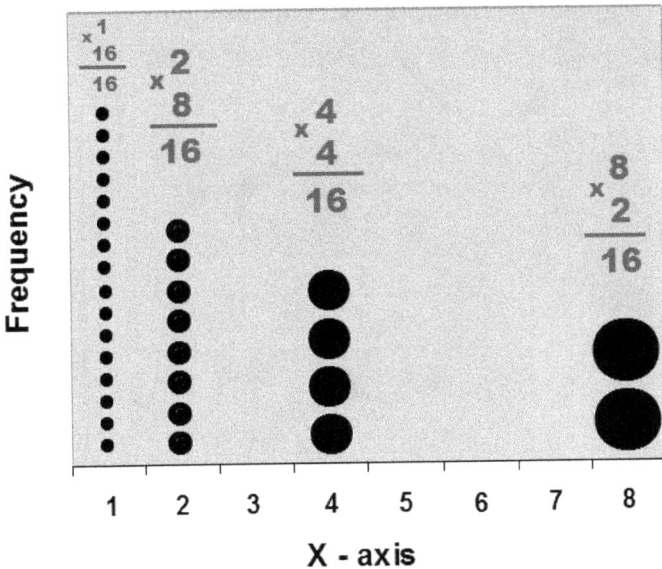

Figure 3.13: Total Quantity or (Frequency) × (Size) Is Constant Throughout for 16/X Distribution

(Frequency)(Size) = (2)(8) = 16; total quantity for size 4 is (4)(4) = 16; total quantity for size 2 is (8)(2) = 16; and total quantity for size 1 is (16)(1) = 16. Surely bigger sizes might appear as if they contain much more total quantity simply by virtue of being big, but they are also rarer so there are fewer of them. This tradeoff between the size and its frequency perfectly cancels out under the postulate. In other words, there exists a perfect balance between size and its frequency so that:

Total Quantity = (Frequency)(Size) = Constant

It should be emphasized that this feature is found most readily and without the need of aggregation only in Exponential Growth, while the other 14 data sets and cases show marked variations in total quantity per size, developing in a certain way, rising from low sizes in the beginning at the minimum, reaching a plateau somewhere before the middle part, then falling monotonically until the end at the maximum, hinting in a sense at a constant total quantity per size only in the aggregate from the minimum to the maximum.

The partitioning of the oval-shaped area in Figure 2.5, having an equitable mix of small, medium, and big, is perfectly in the spirit of the postulate. It is another useful visualization of the conceptual principle behind this quantitative configuration.

Chapter 50

Application of the Postulate
via Generic Bin Schemes on K/X

The aim of this chapter is only to present an overall sketch or outline of how the generic mathematical expression for all possible bin schemes is obtained from the postulate, leaving numerous gaps in the derivations and a lot of missing material, for the sake of brevity. The curious reader is referred to Kossovsky (2014) — the author's first book on Benford's Law and the General Law of Relative Quantities — where a thorough and very detailed exposition is given.

The postulate of the previous chapter shall now be applied in order to arrive at an exact expression of bin proportions for all D and F combinations. This is done via explorations of results from a particular bin scheme having generic D and F values, and performed on the abstract K/X distribution defined from point P to infinity.

In sharp contrast with all the previous empirical bin schemes which start from the 0 origin, this theoretical bin scheme starts from point $P > 0$. It is not possible to start the whole scheme from the origin 0, since $K/0$ is infinite, or rather undefined. Moreover, a very particular P value is chosen here so that $P = W$, namely the same value as that of the width of the bins in the first cycle. It is necessary to establish equality between the width W of the first cycle, and the separation P of the entire system of infinite histograms from the 0 origin.

Figure 3.14 depicts the first cycle of a particular example of such an arrangement of histograms with four-bin scheme on K/X. Here the grey lines are shown in order to separate the bins.

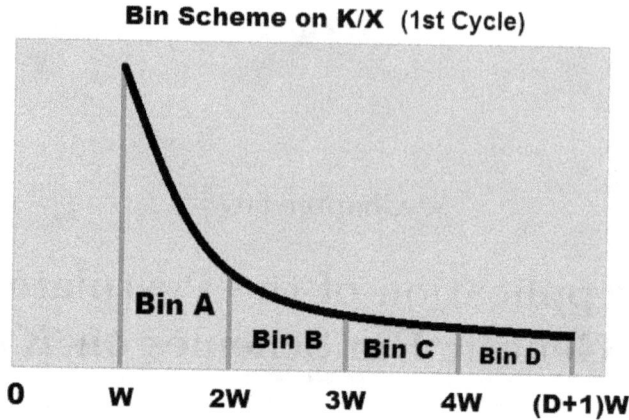

Figure 3.14:   First Cycle of an Infinite Four-Bin Scheme Starting from W

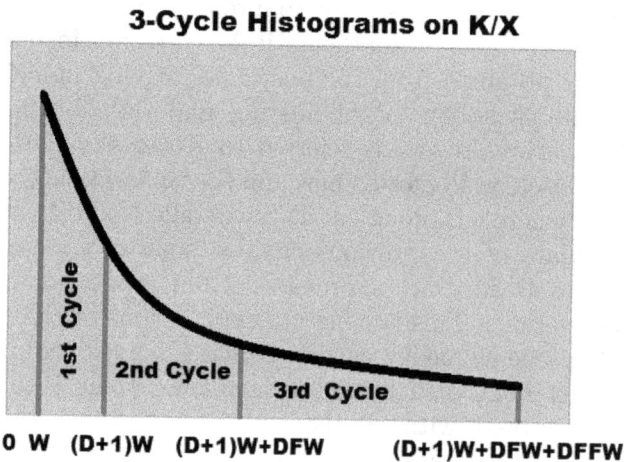

Figure 3.15:   Three Cycles of an Infinite Four-Bin Scheme with $F = 2$

Figure 3.15 depicts the initial three cycles of this particular example of a four-bin scheme on K/X with 2 as the value of inflation factor F. For the sake of brevity and for better visualization, the grey lines separating the bins within each cycle are not shown in Figure 3.15, rather here the grey lines are shown in order to separate the cycles themselves. The inflation factor of 2 implies that bin widths are doubling with each new cycle, as can be seen in Figure 3.15.

The particular choice of $D = 4$ and $F = 2$ for Figures 3.14 and 3.15 is for illustration purposes only, and it does not restrict the generality of subsequent results. This is so since the expressions for the various crucial points on the x-axis (i.e. limits of integrations) in Figures 3.14 and 3.15 are written in extreme generality, based on the generic variables D and F. In additions, the subsequent expressions for the definite integrals are based on the generic variables D and F as well, as opposed to the particular $D = 4$ and $F = 2$ exemplary values.

For the (theoretical) K/X bin scheme, areas as probabilistic proportions are measured here. This is in sharp contrast with all previous (empirical) bin schemes where counts of discrete data points falling within bins were measured.

Let us now evaluate probabilistic areas of K/X for the generic bin scheme constructed with any D and F combinations. This involves evaluating definite integrals over the relevant sub-ranges. In the calculations of 1 cycle, 2 cycles, 3 cycles, and so forth, the span of the range of K/X gradually increases to fit more and more cycles, hence the value of K changes accordingly in order to ensure that total area under the curve is 1. In other words, we are not evaluating distinct parts of a fixed infinite scheme with K approaching 0 in the limit, but rather we are evaluating distinct whole bin schemes of finite ranges which grow in scope with each successive evaluation.

Here lower case d represents bin rank value, so that $d \in \{1, 2, 3, \ldots, D\}$.

$d = 1$ represents Bin A

$d = 2$ represents Bin B

$d = 3$ represents Bin C, and so forth.

The result for one-cycle scheme is as follows:

Proportion(d) $= \ln(1 + 1/d)/\ln(1 + D)$ and can be written as

$\ln([1 + (d)]/[1 + (d - 1)])/\ln(1 + D)$

The result for two-cycle scheme is as follows:

Proportion(d) $= [\ln([1 + (d)]/[1 + (d - 1)])$

$+ \ln([1 + D + (d)F]/[1 + D + (d - 1)F])]/\ln(1 + D + DF)$

For three-cycle scheme, the relevant three definite integrals and their evaluations are:

$$\int_{dW}^{(d+1)W} \frac{K}{x}\,dx + \int_{(D+1)W+(d-1)FW}^{(D+1)W+(d)FW} \frac{K}{x}\,dx$$

$$+ \int_{(D+1)W+DFW+(d-1)FFW}^{(D+1)W+DFW+(d)FFW} \frac{K}{x}\,dx$$

$$\frac{\ln\left(\frac{[1+(d)]}{[1+(d-1)]}\right) + \ln\left(\frac{[1+D+(d)F]}{[1+D+(d-1)F]}\right) + \ln\left(\frac{[1+D+DF+(d)F^2]}{[1+D+DF+(d-1)F^2]}\right)}{\ln(1+D+DF+DF^2)}$$

It is noted that the terms W and K cancel out and drop from the resultant expressions. The cancelation of W is significant as it lends the result a measure of universality rendering it totally independent of the separation between the bin structure and the 0 origin.

One more such exhausting and time-consuming calculation of definite integrals is made in the case of four-cycle scheme constructed for K/X. Luckily there seen to be no need to continue laboring hard again and again, since the resultant algebraic expressions for bin proportions of higher expansion orders follow a repetitive and clear pattern as a long sequence of ever increasing terms in the numerator and ever increasing terms in the denominator. For example, the sequence of the first four denominators is as follows:

$\ln(1 + D)$

$\ln(1 + D + DF)$

$\ln(1 + D + DF + DF^2)$

$\ln(1 + D + DF + DF^2 + DF^3)$

And certainly one may continue blindly building up subsequent terms for the denominator with total confidence; without any additional calculations of definite integrals. The same applies for the terms of the numerator. These quick extrapolations of higher orders expressions enable us to examine numerical examples on the computer for any values of F and D, strongly suggesting that the ratio numerator/denominator reaches a limiting process; and that beyond certain number of initial sequences nothing much is added thereafter. In order to enable us to decipher the eventual limit of the sequence as the number of cycles goes to infinity, the initial four expressions of probabilistic proportions within each bin rank d for one-cycle, two-cycle, three-cycle, and four-cycle schemes, are outlined as follows:

$$\frac{\ln\left(\frac{[1+(d)]}{[1+(d-1)]}\right)}{\ln(1+D)}$$

$$\frac{\ln\left(\frac{[1+(d)]}{[1+(d-1)]}\right) + \ln\left(\frac{[1+D+(d)F]}{[1+D+(d-1)F]}\right)}{\ln(1+D+DF)}$$

$$\frac{\ln\left(\frac{[1+(d)]}{[1+(d-1)]}\right) + \ln\left(\frac{[1+D+(d)F]}{[1+D+(d-1)F]}\right) + \ln\left(\frac{[1+D+DF+(d)F^2]}{[1+D+DF+(d-1)F^2]}\right)}{\ln(1+D+DF+DF^2)}$$

$$\frac{\ln\left(\frac{[1+(d)]}{[1+(d-1)]}\right) + \ln\left(\frac{[1+D+(d)F]}{[1+D+(d-1)F]}\right) + \ln\left(\frac{[1+D+DF+(d)F^2]}{[1+D+DF+(d-1)F^2]}\right) + \ln\left(\frac{[1+D+DF+DF^2+(d)F^3]}{[1+D+DF+DF^2+(d-1)F^3]}\right)}{\ln(1+D+DF+DF^2+DF^3)}$$

The constraint for D is that it must be an integral value greater than 1; namely any positive integer from 2 upwards. This is so because a bin scheme with only 1 bin for each histogram would be meaningless. Since we seek to have a meaningful and proper growth in the sizes of the bins on each consecutive cycle, therefore another constraint here is $F > 1$. The value of F could be any real number, fractional or integral, rational as well as irrational.

Chapter 51

The General Law of Relative Quantities (GLORQ)

Several attempts to obtain the mathematical expression for the limit of the infinite sequence of the previous chapter proved futile, and discussions with several mathematicians about solving this challenging problem did not lead to any progress. Finally, enlisting the help of the distinguished mathematician George Andrews immediately led to fruitful results. George Andrews is the world's leading expert in Integer Partitions, well-known also for his discovery of Ramanujan's lost notebook and his subsequent work on its contents. His reduction of the limit of the infinite sequence to a closed form (analytical) expression — rigorously worked out and detailed in Kossovsky (2014) Chapter 131 — resulted in the following succinct expression:

$$\text{Proportion(d)} = \frac{\ln\left(\dfrac{D + d(F-1)}{D + (d-1)(F-1)}\right)}{\ln(F)}$$

where D is the number of bins, F is the inflation factor, and lower case d is the bin rank.

The notation "ln" refers to the natural logarithm with base e.

The above expression of the limit of the infinite sequence regarding bin schemes constructed for K/X distribution shall be coined as "The General Law of Relative Quantities", and GLORQ as acronym. The adjective "General" is added to indicate that this quantitative phenomenon is the driving force behind the digital phenomenon of Benford's Law; the latter being merely a special case and a consequence of GLORQ, as shall be seen in Section 4. If one chooses to ignore Benford's Law and focus on the generic phenomenon then "The Law of Relative Quantities" would be the more concise and more appropriate name.

The moment of truth has arrived. Does the postulate truly reflect on how relative quantities occur in physical processes and the natural world? Is the complex mathematical analysis correct? Can GLORQ be considered a physical law of nature, or is it merely some mathematical abstraction, totally divorced from the real world? In order to verify that most real-life data sets and physical processes truly follow the above GLORQ expression, empirical results from the upper parts of the tables of Figures 3.7 and 3.8 — regarding those 15 real-life data sets and cases with consistent proportions — shall be examined and compared to the theoretical expression above. Certainly, these empirical confirmations of the theory should apply only corresponding bin systems having identical F and D values for the empirical as well as for the theoretical, in order to obtain relevant comparisons.

The theoretical GLORQ expression shall be evaluated for the two bin schemes of the tables in Figures 3.7 and 3.8 by substituting actual values for the corresponding D and F variables, namely $\{D = 5, F = 2\}$ for Figure 3.7, and separately $\{D = 3, F = 3\}$ for Figure 3.8. Then, for the empirical part, the average of the 15 data sets and cases shall be calculated for each table separately, in order to obtain for each table a singular set of proportions representing the empirical result in its entirety. The theoretical to empirical comparisons are as follows:

General Law of Relative Quantities $D = 5$ $F = 2$
$\{26.3\%, 22.2\%, 19.3\%, 17.0\%, 15.2\%\}$

Empirical Results of Real-life Data $D = 5$ $F = 2$
$\{26.2\%, 22.1\%, 19.2\%, 16.9\%, 15.5\%\}$

General Law of Relative Quantities $D = 3$ $F = 3$
$\{46.5\%, 30.6\%, 22.9\%\}$

Empirical Results of Real-life Data $D = 3$ $F = 3$
$\{46.7\%, 30.5\%, 22.8\%\}$

A remarkable agreement between the theoretical and the empirical is observed and GLORQ is decisively confirmed! Nonetheless, due to the random nature of the law, tiny deviations are seen in each bin scheme as is the case in all statistical laws. Individually, for each data set standing alone, there is still a tiny bit more deviation from GLORQ, but only slightly so, and thus it is deemed totally insignificant, constituting the typical random noise as in all statistical processes.

Figure 3.16 depicts the excellent fit between the theoretical GLORQ proportions and the average of the 15 data sets and cases of Figure 3.7 under the bin scheme $D = 5$ and $F = 2$.

Figure 3.17 depicts the excellent fit between the theoretical GLORQ proportions and the average of the 15 data sets and cases of Figure 3.8 under the bin scheme $D = 3$ and $F = 3$.

The mathematical expression of GLORQ implies the small is beautiful phenomenon, and this feature can be verified by mathematically analyzing the expression and demonstrating that proportion for higher rank bin $(d + 1)$ is always less than proportion for lower rank bin (d). In addition, since GLORQ was obtained by constructing bin schemes for the K/X distribution, it follows that lower rank bins are always of higher proportions in comparison to higher rank bins due to the fact that the curve of the K/X distribution itself is consistently and monotonically falling to the right.

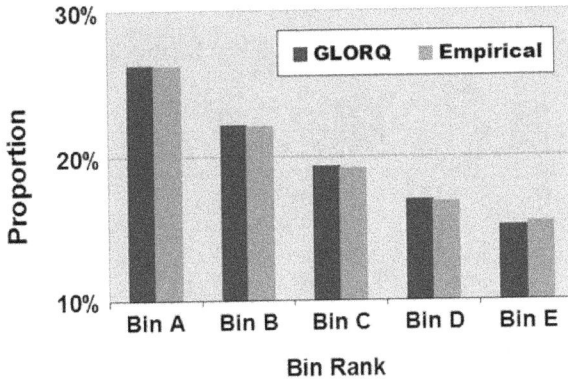

Figure 3.16: Nearly Perfect Agreement between Theoretical and Empirical, $D = 5$, $F = 2$

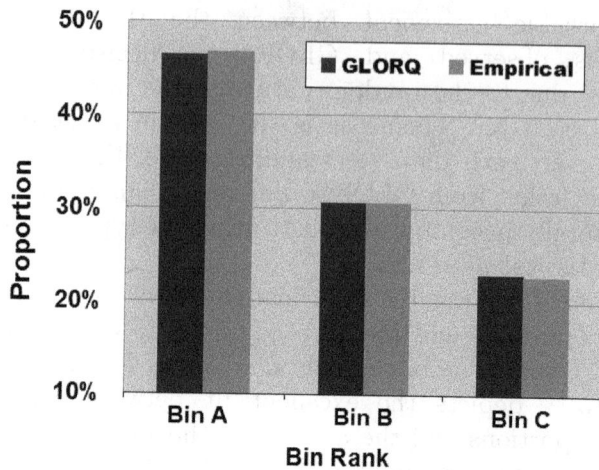

Figure 3.17: Nearly Perfect Agreement between Theoretical and Empirical, $D = 3$, $F = 3$

George Andrews also suggested a straightforward algebraic manipulation of the GLORQ expression in order to unmistakably show that it is inversely proportional to d, thus implying the small is beautiful phenomenon. His derivation follows:

$$\frac{\ln\left(\dfrac{D + d(F - 1)}{D + (d - 1)(F - 1)}\right)}{\ln(F)}$$

Expanding a bit the denominator of the numerator:

$$\frac{\ln\left(\dfrac{D + d(F - 1)}{D + (d)(F - 1) - (F - 1)}\right)}{\ln(F)}$$

Subtracting $(F - 1)$ and adding $(F - 1)$ on top:

$$\frac{\ln\left(\dfrac{D + d(F - 1) - (F - 1) + (F - 1)}{D + (d)(F - 1) - (F - 1)}\right)}{\ln(F)}$$

Further reducing the numerator:

$$\frac{\ln\left(1 + \dfrac{(F-1)}{D + (d)(F-1) - (F-1)}\right)}{\ln(F)}$$

Simplifying the denominator of the numerator:

$$\frac{\ln\left(1 + \dfrac{(F-1)}{D + (d-1)(F-1)}\right)}{\ln(F)}$$

This new expression for GLORQ has the d variable appearing only once in the denominator of the numerator, hence GLORQ is inversely proportional to d, and proportion for higher rank bin $(d+1)$ is always less than proportion for lower rank bin (d).

Needless to say, the near perfect agreement between theoretical GLORQ and empirical results is not only encountered in the cases of $\{D = 5, F = 2\}$ and $\{D = 3, F = 3\}$ bin schemes, but in all bin schemes of any D and F values whatsoever. A good illustration of the compatibility between theoretical and empirical results here is given with regard to the variety of empirical bin schemes performed on the Earthquakes data set as shown in Figure 3.11. Applying the same combinations of D and F values for the theoretical GLORQ expression yields the result shown in Figure 3.18. The tables in Figures 3.11 and 3.18 are very much alike, with only very mild differences in values. The average of the absolute values of the 74 differences between the tables is merely 0.41%. Had more effort been put here in obtaining the average of all the 15 data sets and cases, instead of the particular data on earthquakes, then even better agreement would have been seen between theoretical and empirical results.

D	F	Bin A	Bin B	Bin C	Bin D	Bin E	Bin F	Bin G
3	2	41.5%	32.2%	26.3%				
3	3	46.5%	30.6%	22.9%				
3	4	50.0%	29.2%	20.8%				
3	5	52.6%	28.1%	19.3%				
3	6	54.7%	27.1%	18.2%				
3	11	61.2%	23.8%	15.1%				
4	2	32.2%	26.3%	22.2%	19.3%			
4	3	36.9%	26.2%	20.3%	16.6%			
4	4	40.4%	25.7%	18.9%	15.0%			
4	5	43.1%	25.2%	17.9%	13.9%			
4	12	53.2%	22.1%	14.2%	10.5%			
5	2	26.3%	22.2%	19.3%	17.0%	15.2%		
5	5	36.5%	22.8%	16.7%	13.1%	10.8%		
5	9	43.5%	21.8%	14.7%	11.1%	8.9%		
7	2	19.3%	17.0%	15.2%	13.8%	12.6%	11.5%	10.7%
7	4	25.7%	18.9%	15.0%	12.4%	10.6%	9.2%	8.2%
7	8	33.3%	19.5%	13.8%	10.7%	8.8%	7.4%	6.4%

Figure 3.18: A Variety of Bin Schemes with Selected D and F Values — Theoretical GLORQ

Chapter 52

Saturation in GLORQ-Inducing Processes Precludes Extreme Skewness

Three generic causes of the small is beautiful phenomenon were outlined in Section 2:

Partition processes.
Multiplication processes.
Data aggregations and the related model of the chains of distributions.

These processes take non-skewed data sets as input and turn them into skewed data as output. Often this is done gradually, in a cumulative way, applying the process over and over again.

Crucially, and as if by magic, when these processes are repeatedly applied numerous times, they do not lead to extreme skewness in resultant data where the small overwhelms all other sizes, and certainly not to the total elimination of the big and the medium. Rather, saturation in quantitative transformation eventually occurs leading to the GLORQ quantitative configuration, and no further movement is made towards additional advancement of the small. These three processes tend to strongly reconfigure input data with additional skewness only initially, but their ability to do so diminishes as the GLORQ quantitative configuration is nearly achieved, and finally these processes have no further quantitative effect on input data beyond that point. Miraculously, they all stop at GLORQ!

Multiplication Process	Bin A	Bin B	Bin C	Bin D	Bin E
1 Ten-Sided Die	19.9%	19.8%	9.9%	30.0%	20.4%
2 Ten-Sided Dice	24.7%	21.1%	20.2%	17.0%	17.0%
3 Ten-Sided Dice	26.7%	19.9%	22.5%	13.2%	17.7%
4 Ten-Sided Dice	27.1%	19.6%	22.3%	13.9%	17.1%
5 Ten-Sided Dice	27.4%	19.9%	21.8%	15.9%	15.1%
6 Ten-Sided Dice	26.7%	21.0%	20.2%	16.6%	15.5%
7 Ten-Sided Dice	26.8%	21.6%	19.4%	17.0%	15.2%
14 Ten-Sided Dice	26.2%	22.6%	19.3%	16.8%	15.2%
GLORQ	26.3%	22.2%	19.3%	17.0%	15.2%

Figure 3.19:   Saturation in Multiplication Processes Leading to GLORQ

As a demonstration of this saturation feature in at least one of the above processes, several multiplication processes involving 10-sided dice are studied via 10,000 Monte Carlo computer simulations. The table in Figure 3.19 pertains to results from empirical bin schemes applied for each of these random multiplication processes. The first row is actually not about any multiplication process, but rather about a single die with 10 sides $\{1, 2, 3, 4, 5, 6, 7, 8, 9, 10\}$, randomly thrown 10,000 times, and values of the faces recorded. The second row is about two dice, each with 10 sides, both randomly thrown simultaneously 10,000 times, recording the product on each throw. The next to the last row is about 14 dice, each with ten sides, all randomly thrown simultaneously 10,000 times, recording the product on each throw. The parameters in all of these bin schemes are $D = 5$, $F = 2$, and $W = 0.03$. The last row is of the theoretical GLORQ proportion for $D = 5$, $F = 2$. As can be seen in Figure 3.19, results gradually converge to the GLORQ proportions, and finally saturation emerges at the end. Proportions never surpass or overrun the GLORQ limit; no matter how many times multiplication is being applied. Here for the 10-sided dice, it takes about 14 or so such random multiplication processes to arrive approximately at the GLORQ configuration, achieving quantitative stability, and no changes occur beyond this point. Figure 3.20 converts the numerical table in Figure 3.19 into bar charts, helping to visualize these saturation effects and the eventual convergence to GLORQ.

Multiplications of 10-Sided Dice

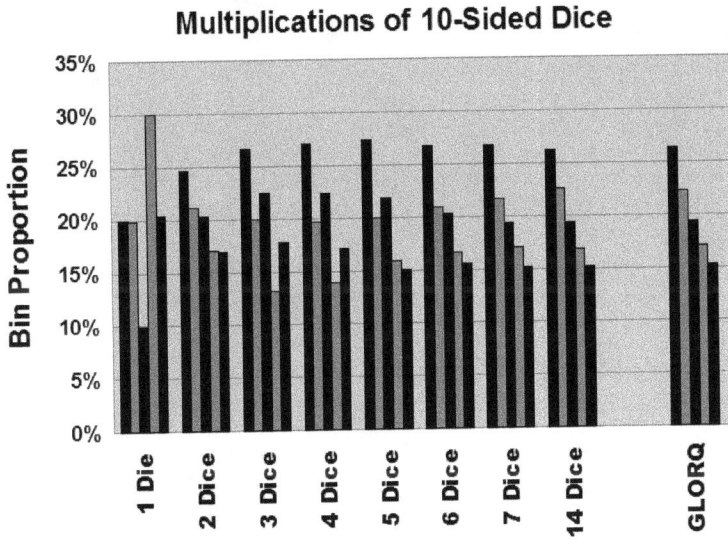

Figure 3.20:   Visualization of Saturation in Multiplications Leading to GLORQ

The same saturation effects are found in partition processes, as well as in data aggregations and chains of distributions. For example, 10,000 pipes — each of random length chosen from the continuous Uniform Distribution between 5 meters and 15 meters — are partitioned gradually, in stages, by placing random marks on each pipe via the Uniform Distribution. Initially, only 1 mark is placed on each pipe, resulting in 20,000 pipes. Then another mark is placed resulting in 2 marks total, and 30,000 pipes. Then another mark is placed resulting in 3 marks total, and 40,000 pipes, and so forth. Empirical Monte Carlo computer simulations show that this leads very gradually to GLORQ and that saturation finally sets in preventing the process from surpassing the GLORQ configuration limit.

In another example, a long chain of Uniform Distributions is assembled, gradually, by adding additional parametrical dependencies in stages. The first chain is Uniform(0, Uniform(0, R)). The second chain is Uniform(0, Uniform(0, Uniform(0, R))). The third chain is a sequence of the combination of four Uniforms as in Uniform(0, Uniform(0, Uniform(0, Uniform(0, R)))), and so forth. Here very rapid saturation and convergence to GLORQ is seen in Monte Carlo computer simulations, and which never surpasses that GLORQ configuration limit.

Chapter 53

GLORQ is Number System Invariant

The small is beautiful phenomenon is an obvious consequence of the General Law of Relative Quantities. The question then naturally arises whether GLORQ is a universal, quantitative, and physical measure, or simply a numerical consequence of our arbitrarily invented number system. Our positional number system was completed in the Renaissance Period after the introduction of the 0 digit by Aryabhatta and Brahmagupta in India in the 5th and 6th centuries and the work of the 8th century Arabic mathematician Al-Khwarizmi, and it is extremely efficient, yet it is arbitrary, an invented system which facilitates the counting and calculating of quantities. If GLORQ depends on our number system, then it should be considered arbitrary as well.

In order to answer the question regarding the universality of GLORQ, one needs to determine whether or not histograms depend in any way on the number system in use. In other words, does the visual structure and the shape of a histogram change when we apply say base 4 number system instead of the usual base 10, or when we switch to another number system altogether such as Roman numerals? This is the crux of the matter. If histograms get distorted and change whenever the number system is switched, then there is very little hope that GLORQ is universal, but if histograms are immune to these number system changes then GLORQ is surely a universal law. This is so since GLORQ is built and based on the structure of the histogram of the data set under consideration.

The well-known planet Vizubla located deep within the Orion Arm of the Milky Way Galaxy is without an atmosphere; hence it is clearly visible (as suggested by the name Vizubla) throughout the galaxy. Vizubla is often used by scientists and statisticians throughout the galaxy to verify certain physical laws and statistical patterns. The particular data chosen to be examined in the context of GLORQ is the length of all the rivers in Vizubla. Relatively not too far away, in the same vicinity of the Orion Arm, there exist four different planets with intelligent civilizations, each with its own unique number system. Two planets use positional number systems with base 10 and with base 4; the third planet uses Roman numerals; and the fourth planet has never invented a number system, as it simply uses a very large set of distinct symbols to denote quantities.

Vizubla's river phenomenon is carefully observed and meticulously recorded on these four planets, and then compatibility with GLORQ is checked in order to verify the validity of the law. Since each planet uses its own unique number system, could it then be inferred that these four histograms should appear very differently? Clearly, each data point relating to a particular river in Vizubla is recorded in a very different manner numerically on these four planets, and all this could possibly point to profoundly different histogram drawings. Consequently, this could preclude any unified GLORQ-like law relevant to all observers.

Let us now demonstrate that histogram construction is actually number system invariant, and that these four histograms should appear with the same structure and shape, in spite of the profoundly distinct ways the planets numerically record observed quantities. Figures 3.21–3.24 depict the four histograms drawn on these four distinct planets regarding Vizubla river data.

A moment thought would convince anyone that these four histograms are visually identical, except for the designation of the various numbers and symbols below the horizontal axis and to the left of the vertical axis.

In fact, within the horizontal axis as well as within the vertical axis, each value represents a primitive count which does not necessitate any number system. That count is geometrical, where 1 unit of distance along the squares represents a certain quantity, irrespective of the number system in use. The horizontal axis represents the scale of the measured quantity, while the vertical axis

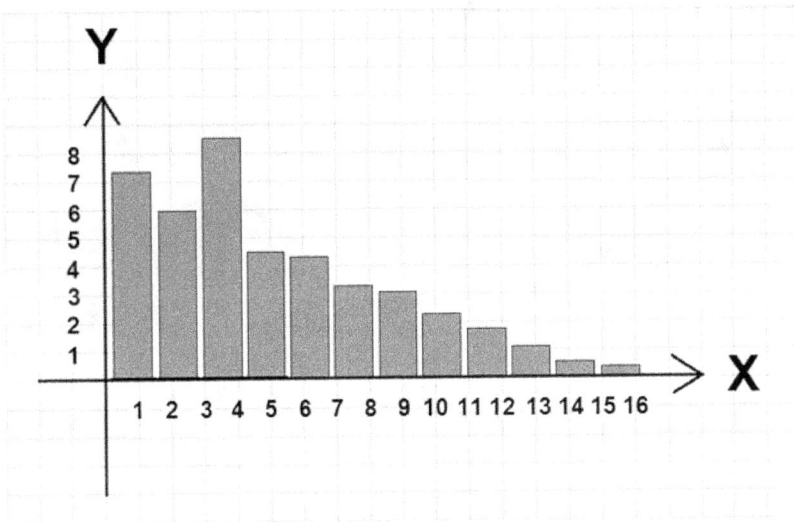

Figure 3.21: Histogram of Vizubla River Data with Positional Number System Base 10

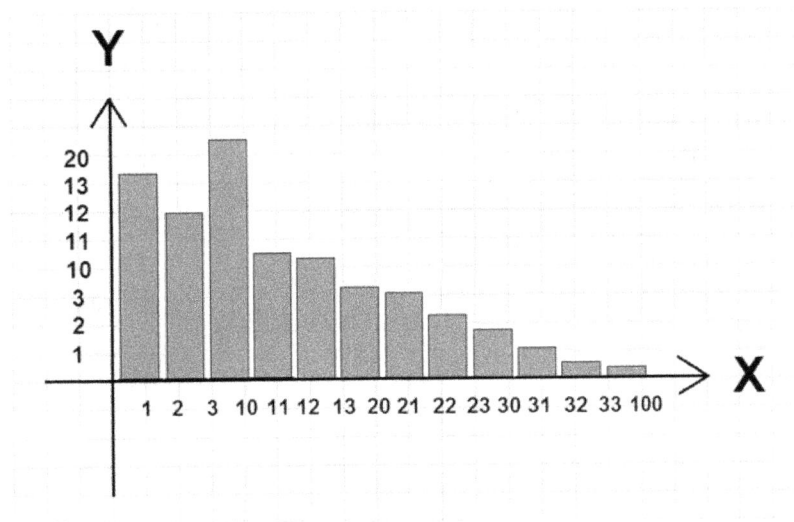

Figure 3.22: Histogram of Vizubla River Data with Positional Number System Base 4

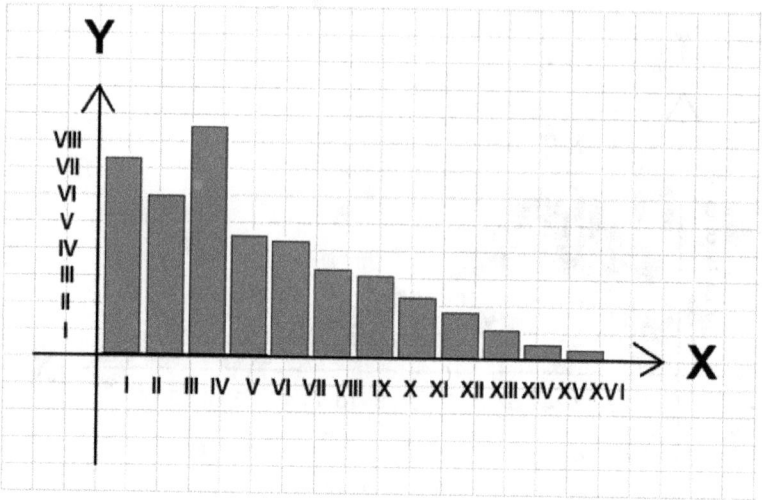

Figure 3.23:   Histogram of Vizubla River Data with Roman Numerals

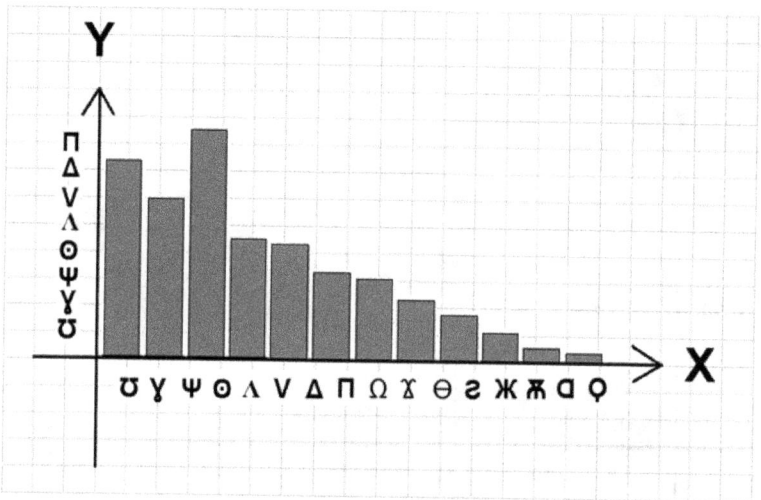

Figure 3.24:   Histogram of Vizubla River Data Absent a Number System

represents the count of how many data points fall within the bin of a certain sub-range of the scale. Hence, the visual message conveyed in a given histogram is universal and independent of societal number systems.

The fact that the shape of these four histograms in Figures 3.21–3.24 is universal and agreed on by all four planets, strongly suggests that there exists one basic and universal statistical measure that could be verified by all of them. Surely, that measure is GLORQ!

In the same vein, all four planets can directly and visually observe at the raw level that there are numerous small rivers, some medium-sized rivers, but very few big rivers — and surely this observation is universal. They can also directly observe the exact sizes and relative quantities of these rivers and could perhaps discover a certain statistical pattern in the data, and all these observations are physical, real, and objective, absolutely independent of any number system in use, or even histogram construction. Since these rivers appear identically in all four planets and are universally observed, it follows that there exists a singular quantitative pattern for them stated as a physical law which is true for all observers, regardless of the observer's number system. Surely, that measure is GLORQ!

Since statistical distributions (probability density functions) are simply the continuous forms of discrete histograms (i.e. infinity-refined histograms), it follows that the histogram invariance principle is even more general, and all probability density distributions are number system invariant as well.

SECTION 4

Benford's Law as a Direct Consequence of the General Law

Chapter 54

The Physical Clues Leading to the Discovery of Benford's Law

Benford's Law describes how digits are spread within numbers of real-life data. The law states that low digits such as 1, 2, and 3 occur more frequently than high digits such as 7, 8, and 9. Benford's Law was discovered by an indirect inference about patterns in occurrences of digits within numbers of actual data. The two discoverers were Simon Newcomb in 1881, and (independently) Frank Benford in 1938. What caught their attention was the observation about the differentiated physical wear and tear of the pages in old books of tables of logarithms — commonly used by engineers and scientists before the advent of calculators and computers. These old books seemed to be more strained by use and quite worn in the first pages relating to first digits 1, 2, and 3, and progressively less so throughout the book for higher digits, culminating in the last pages relating to first digits 7, 8, and 9 which seemed to be in relatively excellent condition, as if they haven't been much in use.

Naturally, Newcomb and Benford inferred that the engineers and scientists reading these logarithmic books were overall more in need of using the first pages of low [first] digits than the last pages of high [first] digits, reflecting the spread of the [first] digits in the real world; namely that such differentiated use in the pages for real-life numbers in actual data indirectly indicates that numbers beginning with low digits occur quite frequently, while numbers beginning with high digits occur less frequently.

Newcomb's short two-page article begins by mentioning this differentiation in the usage of the pages. *That the ten digits do not occur with equal frequency must be evident to any one making much use of logarithmic tables, and noticing how much faster the first pages wear out than the last ones.*

Benford's elaborated article begins with a discussion about what this differentiation in the physical wear and tear of old logarithmic books may ultimately hint at. *It has been observed that the pages of a much used table of common logarithms show evidences of a selective use of the natural numbers. The pages containing the logarithms of the low numbers 1 and 2 are apt to be more strained and frayed by use than those of the higher numbers 8 and 9. Of course, no one could be expected to be greatly interested in the condition of a table of logarithms, but the matter may be considered more worthy of study when we recall that the table is used in the building up of our scientific, engineering, and general factual literature.*

Chapter 55

The First Digit on the Left Side of Numbers

Newcomb and Benford have discovered that the first digit on the left-most side of numbers in real-life data sets is most commonly of low value such as $\{1, 2, 3\}$ and rarely of high value such as $\{7, 8, 9\}$. As an example serving as a brief and informal empirical test, a sample of 40 values relating to geological data on time between earthquakes is randomly chosen from the data set on all global earthquake occurrences in 2012 — in units of seconds. Figure 4.1 depicts this sample of 40 numbers. Figure 4.2 emphasizes in bold and black color the first digits of these 40 numbers.

Clearly, for this very small sample, low digits occur by far more frequently on the first position than do high digits. A summary of the digital configuration of the sample is given as follows:

Digit Index:	$\{1, 2, 3, 4, 5, 6, 7, 8, 9\}$
Digits Count totaling 40 values:	$\{15, 8, 6, 4, 4, 0, 2, 1, 0\}$
Proportions of Digits with "%" sign omitted:	$\{38, 20, 15, 10, 10, 0, 5, 3, 0\}$

Assuming (correctly) that these 40 values were collected in a truly random fashion from the large data set of all 19,452 earthquakes occurrences in 2012; without any bias or attempt to influence first digits occurrences; and that this pattern is generally found in many other data sets, one then may conclude with the phrase "not all digits are created equal", or rather "not all first digits are created equal",

285.29	185.35	2579.80	27.11
5330.22	1504.49	1764.41	574.46
1722.16	815.06	3686.84	1501.61
494.17	362.48	1388.13	1817.27
3516.80	5049.66	2414.06	387.78
4385.23	2443.98	2204.12	1224.42
1965.46	3.61	1347.30	271.23
3247.99	753.80	1781.45	593.59
1482.64	1165.04	4647.39	1219.19
251.12	7345.52	1368.79	4112.13

Figure 4.1:   Sample of 40 Time Intervals between Earthquakes

285.29	185.35	2579.80	27.11
5330.22	1504.49	1764.41	574.46
1722.16	815.06	3686.84	1501.61
494.17	362.48	1388.13	1817.27
3516.80	5049.66	2414.06	387.78
4385.23	2443.98	2204.12	1224.42
1965.46	3.61	1347.30	271.23
3247.99	753.80	1781.45	593.59
1482.64	1165.04	4647.39	1219.19
251.12	7345.52	1368.79	4112.13

Figure 4.2:   The First Digits of the Earthquake Sample

even though this seems to be contrary to intuition and against all common sense.

The focus here is actually on the *first meaningful digit* — counting from the left side of numbers, excluding any possible encounters of zero digits which only signify ignored exponents in the relevant set of powers of 10 of our number system. Therefore, the complete definition of the "First Leading Digit" is the *first non-zero digit* of any given number on its left-most side. This digit is the first significant one in the number as focus moves from the left-most position towards the right, encountering the first non-zero digit signifying some quantity; hence it is also called the "First Significant Digit". For 2365 the first leading digit is 2. For 0.00913 the first leading digit is 9 and the zeros are discarded; hence even though strictly speaking the first digit on the left-most side of 0.00913 is 0, yet, the first significant digit is 9. For the lone integer 8 the leading digit is simply 8. For negative numbers the negative sign is discarded, hence for -715.9 the leading digit is 7. Here are some more illustrative examples:

6,719,525	$\rightarrow$	digit 6
0.0000761	$\rightarrow$	digit 7
-0.281264	$\rightarrow$	digit 2
875	$\rightarrow$	digit 8
3	$\rightarrow$	digit 3
-5	$\rightarrow$	digit 5

For a data set where all the values are greater than or equal to 1, such as in the sample of the earthquake data, the first digit on the left-most side of numbers is also the First Leading Digit and the First Significant Digit, and necessarily one of the nine digits $\{1, 2, 3, 4, 5, 6, 7, 8, 9\}$; while digit 0 never occurs first on the left-most side.

Chapter 56

Benford's Law and the Predominance of Low Digits

Benford's Law states that:

Probability[First Leading Digit is d] = $\mathbf{LOG_{10}(1 + 1/d)}$

$LOG_{10}(1 + 1/1) = LOG(2.00) = 0.301$
$LOG_{10}(1 + 1/2) = LOG(1.50) = 0.176$
$LOG_{10}(1 + 1/3) = LOG(1.33) = 0.125$
$LOG_{10}(1 + 1/4) = LOG(1.25) = 0.097$
$LOG_{10}(1 + 1/5) = LOG(1.20) = 0.079$
$LOG_{10}(1 + 1/6) = LOG(1.17) = 0.067$
$LOG_{10}(1 + 1/7) = LOG(1.14) = 0.058$
$LOG_{10}(1 + 1/8) = LOG(1.13) = 0.051$
$LOG_{10}(1 + 1/9) = LOG(1.11) = 0.046$
$-----$
1.000

Figure 4.3 depicts the distribution. Figure 4.4 visually depicts Benford's Law as a bar chart.

This set of nine proportions of Benford's Law is sometimes referred to in the literature as "The Logarithmic Distribution". Remarkably, Benford's Law is confirmed in almost all real-life data sets with high order of magnitude, such as in data relating to physics, chemistry, astronomy, economics, finance, accounting, geology, biology, engineering, governmental census data, and many others.

Digit	Probability
1	30.1%
2	17.6%
3	12.5%
4	9.7%
5	7.9%
6	6.7%
7	5.8%
8	5.1%
9	4.6%

Figure 4.3: Benford's Law for First Digits

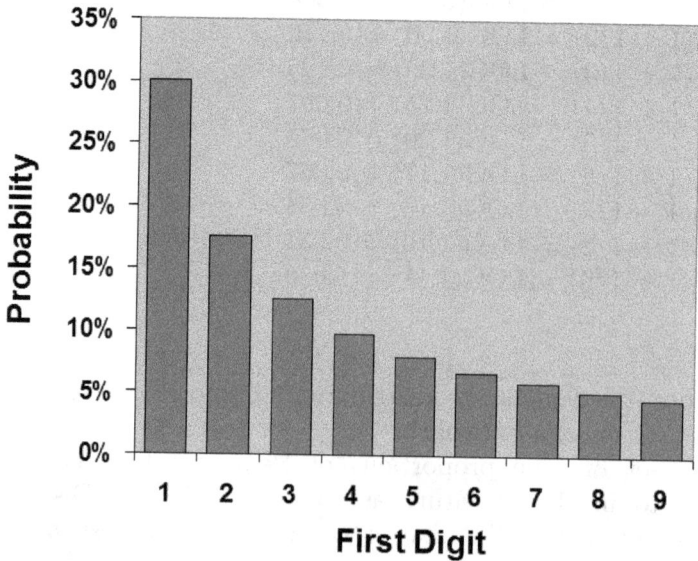

Figure 4.4: Benford's Law — Probability of First Leading Digit Occurrences

A useful expression yielding the first leading digit of any positive number X is given by

$$\text{First Digit of } X = \text{INT}(X/10^{\text{INT}(\text{LOG } X)})$$

The INT function refers to the integer just below X, or to X itself if X is exactly an integer.

For example, $\text{INT}(7.2)$ is 7, $\text{INT}(9)$ is 9, $\text{INT}(0.85)$ is 0, and $\text{INT}(-5.2)$ is -6. The LOG function refers to the decimal logarithm of X with base 10. This expression should prove quite useful in computer implementations of Benford's Law, and especially in MS-Excel which uses the identical names of LOG and INT for these two functions.

Benford's Law is stated purely in terms of the proportions of the digits within numbers in real-life and physical data sets, and as such it is based and depends on our number system. Indeed, there exists a (misguided and mistaken) school of thought that holds that Benford's Law is merely a consequence of our invented number system; that it is a numerical phenomenon, not a physical phenomenon. In other words, that the law reflects aspects and features of our positional number system; and that it is not a law of nature expressing physical and natural reality. As shall be seen in the next few chapters, this vista of Benford's Law is wrong.

Chapter 57

Empirical Tests and Confirmation of Compliance with Benford's Law

Figure 4.5 provides the first digit distributions for the 24 data sets and cases of Chapter 45. In addition, Benford's Law (the Logarithmic Distribution) is shown at the bottom in the last row for comparison. As it happened, nearly every item of the upper class of 15 data sets and cases obeys Benford's Law, while the majority of the items of the lower class of 9 data sets and cases disobey. Hence the same division into 15 and 9 sets which was observed for GLORQ in Figures 3.7 and 3.8 applies to Benford's Law as well, as seen here in Figure 4.5.

In other words, those data sets which obey GLORQ also obey Benford's Law, and those which disobey GLORQ also disobey Benford's Law. Such strong correspondence in compliance between the two laws is a very good indication that GLORQ and Benford's Law might have some very profound connection, where one possibly mirrors the other.

But aren't these two laws fundamentally different?! GLORQ is stated in terms of pure quantities and it is independent of any number system in use, while Benford's Law is stated in terms of the symbolic digits written to express numbers in our base 10 positional number system!? Is it just a miraculous and very rare coincidence, or perhaps there is some deep-seated relationship between these seemingly very different laws. The surprising connection between Benford's Law and GLORQ shall be discussed in chapter 62, where Benford's Law is shown to be simply a special case and a consequence of GLORQ, and this fact nicely explains the correspondence in compliance seen here.

Data Set	1	2	3	4	5	6	7	8	9
Exoplanets	29.7%	19.2%	10.3%	8.6%	8.2%	8.6%	5.1%	4.8%	5.5%
Capitalization	29.6%	16.3%	12.5%	9.3%	8.6%	7.2%	6.3%	5.9%	4.4%
Population	29.4%	18.1%	12.0%	9.5%	8.0%	7.0%	6.0%	5.3%	4.6%
Molecules	31.9%	25.2%	16.1%	8.4%	5.7%	4.3%	2.9%	3.2%	2.3%
Pulsars	33.1%	21.1%	12.7%	8.5%	5.9%	4.8%	5.0%	4.5%	4.4%
Expenses	29.7%	17.7%	12.1%	9.7%	8.6%	6.6%	6.1%	4.9%	4.5%
Price List	28.6%	19.0%	12.3%	9.4%	8.5%	7.1%	6.3%	4.2%	4.6%
House Number	30.9%	18.5%	15.1%	10.0%	6.1%	6.0%	5.2%	4.3%	4.0%
Rock Breaking	29.6%	17.6%	12.6%	9.7%	7.7%	6.6%	6.0%	5.6%	4.5%
Pipe Breaking	27.9%	18.3%	13.1%	10.2%	8.8%	6.9%	5.3%	5.0%	4.5%
Exponential Gr.	29.5%	17.1%	12.1%	9.7%	8.0%	6.2%	6.5%	5.9%	5.0%
Long Chain	30.6%	17.5%	11.8%	9.3%	8.4%	7.0%	5.9%	5.1%	4.5%
Lognormal	31.0%	18.3%	12.3%	9.5%	7.4%	6.6%	5.5%	4.9%	4.6%
Exponential	26.7%	17.3%	13.5%	10.9%	8.2%	7.8%	6.5%	4.8%	4.4%
Earthquakes	29.9%	18.8%	13.5%	9.3%	7.5%	6.2%	5.8%	4.8%	4.2%
Wars	32.1%	16.4%	14.2%	7.5%	9.0%	4.5%	4.5%	4.5%	7.5%
Rivers	65.7%	19.3%	7.2%	3.9%	1.7%	2.2%	0.0%	0.0%	0.0%
Carbon Dioxide	26.9%	16.7%	10.2%	16.2%	9.3%	6.5%	6.5%	4.2%	3.7%
Short Chain	41.3%	9.4%	7.0%	7.1%	7.3%	7.0%	7.4%	7.0%	6.4%
30-Dice Depend.	40.7%	18.4%	7.9%	7.0%	6.1%	5.7%	5.0%	4.8%	4.4%
4-Dice Products	29.2%	16.4%	14.0%	11.4%	4.2%	7.9%	6.8%	3.5%	6.5%
Primes	12.4%	11.8%	11.4%	11.1%	11.0%	10.6%	10.7%	10.5%	10.5%
Uniform	0.0%	32.8%	33.2%	34.1%	0.0%	0.0%	0.0%	0.0%	0.0%
Normal	73.4%	22.5%	0.1%	0.1%	0.1%	0.3%	0.6%	1.0%	1.8%
BENFORD'S LAW	**30.1%**	**17.6%**	**12.5%**	**9.7%**	**7.9%**	**6.7%**	**5.8%**	**5.1%**	**4.6%**

Figure 4.5: Results for 24 Data Sets and Cases — First Digits Proportions and Benford's Law

Actually, Molecules, Pulsars, and Exponential, of the upper class group, deviate a bit from the Logarithmic Distribution, but not significantly so. And actually, Wars, Carbon Dioxide, and the Four-Dice Products, of the lower class group, are somewhat close to the Logarithmic Distribution.

Chapter 58

The Theoretical Clues Leading from Benford's Law to GLORQ

Benford's Law was not discovered by somebody who sat down deep in thought, inquiring about the supposed distribution of digits within numbers in real-life random data, attempting to arrive at an exact mathematical expression in the abstract. Most likely nobody ever contemplated the issue, and almost certainly nobody ever actually examined digit distributions within data before it was discovered by Newcomb and Benford. Rather, the law "forced itself upon us" by gently flaunting its physical manifestation, giving us the clue to the very existence of the phenomenon via the differentiated physical wear and tear of pages in old books of tables of logarithms.

How was GLORQ discovered? Why would anyone conceive of such peculiar bin algorithm as in the structure of GLORQ which miraculously yields nearly universal quantitative measures across an extremely wide variety of data sets and physical processes? The following line of thought prompted the author to generalize the phenomenon from digits to quantities:

(1) Our positional number system, completed around the Renaissance Era with the introduction of the 0 digit is extremely efficient. But it's still arbitrary!

Hence Benford's Law, being so intimately involved with our number system to the extent that it is stated in terms of its symbols (digits), is arbitrary just as well! This realization leads one to suspect that $LOG_{10}(1 + 1/d)$ for the first digits does not account for the full story of the phenomenon, and that there exists possibly a more universal and non-arbitrary law.

It is necessary to remind ourselves of the distinction between the verbs "discover" and "invent".

We have *discovered* that the sides of a right triangle relates as in $C^2 = A^2 + B^2$.

We have *discovered* that $F = M \times A$ and $F_G = G \times M_1 \times M_2/R^2$.

We have *invented* a very efficient number system, the positional number system base 10.

For example, **7205.38** is defined as $\mathbf{7} \times 10^3 + \mathbf{2} \times 10^2 + \mathbf{0} \times 10^1 + \mathbf{5} \times 10^0 + \mathbf{3} \times 10^{-1} + \mathbf{8} \times 10^{-2}$.

Indeed, our numbers are defined as linear combinations of rational fractions.

For example, **7205.38** is defined as $\mathbf{7} \times 1000 + \mathbf{2} \times 100 + \mathbf{0} \times 10 + \mathbf{5} \times 1 + \mathbf{3}/10 + \mathbf{8}/100$.

(2) "Physical Reality Versus Digital Perception": Benford's Law is highly prevalent in the physical world, yet this is found by first having quantities recorded in our number system, then first digits are counted, and finally $\text{LOG}_{\mathbf{10}}(1+1/d)$ is found. Hence in this sense our digits serve as a lens of sorts in viewing the physical phenomenon.

Two radically different interpretations of the Benford phenomenon are given:

First: This is truly a physical phenomenon existing independently of us and our way of recording data, thus it is a physical law of nature.

Second: This digital pattern found in physical data is simply due to our own peculiar way of counting values by way of their digital representations; hence the phenomenon has no independent physical existence outside our digital perception.

As an analogy for the second interpretation, a child wearing red eyeglasses may believe that every physical object in the world is red, and ask his or her father "Daddy, how come everything in the world is red?" The red color on his or her eye glasses is arbitrary, and that's why the fact that everything appears red is arbitrary as well. Had he or she been wearing green eye glasses, everything then would have appeared green.

(3) Do histograms depend on the number system in use? Does the visual picture of a histogram change when we switch to another number system? The surprising answer is that there are no changes! A moment thought would convince everyone that the histogram of a singular real-life data set viewed through the prism of several distinct number systems appear identical. The cartesian plane is in essence a geometrical one, where each unit of length signifies one unit of the quantity being measured, and therefore, the visual aspect of an histogram, the relative sizes of the bins, its shape, and so forth, are all fixed and invariant with respect to the number system in use; well, except for the symbols, digits, numerals, and numbers written along the horizontal and vertical axes, which are indeed number system dependent. Clearly, the message conveyed in a given histogram is universal; irrespective of the number system is use, hence: "Histogram Number-System Invariance Principle". Since statistical density distributions are simply the continuous forms of discrete histograms (infinity refined), the principle is very general, and density distributions are number-system invariant as well.

(4) Imagine two observers on two distinct planets A and B, both observing six distinct quantitative phenomena of a third planet C via powerful telescopes. Planet A has invented a positional number system base 10, and thus very clearly observes Benford's Law where first digits are nearly as in $LOG_{10}(1 + 1/d)$ for all six data sets of planet C. Planet B has invented Roman numerals and is numero-conservative; adamantly refusing to tolerate any other numerical system for cultural and nationalistic reasons. Planet B proclaims that it does not see any unifying pattern in those six data sets of Planet C, and it is asking for help in constructing a quantitative and number-neutral measure that would yield a uniform pattern for all six data sets.

    Philosophically, and according to the first [and correct] interpretations of the Benford phenomenon, it is necessary that both planets should come up with a universal and primitive statistical measure agreed by all for this clearly and easily observable physical phenomenon on planet C, since the phenomenon exists in its own right independently of any observers out there and independently of any arbitrarily invented number system. In other words, that a singular quantitative statement should be formulated which would be identical for all observers, being number system invariant.

But what measure would it be? Naturally, the idea that comes to mind here is that that universal and primitive measure to be agreed on by all observers could be a mathematical expression relating to the commonly observed histograms of the data sets in question (since it was shown earlier to be of such universal character).

But what aspect of the histograms could be measured which could be common to all six histograms and prove to be the universal pattern?

(5) One characteristic common to all Benford obeying data sets is their skewed histogram, falling on the right for high values. This implies having many small values, but only very few big ones, confirming the motto "Small is Beautiful". This is a universal feature, being number system invariant. Therefore, a precise quantitative measure of such a fall in histograms may perhaps serve as a general law, true for all observers regardless of their number system in use.

Hence let's change the agenda, and instead of focusing on digits, let us focus on histograms of data sets and their quantitative structure preferring the small over the big.

(6) How should we go about measuring that fall in the histogram on the right? As was shown in chapters 44 and 45, one single histogram tailor-made according to the range of each individual data set wouldn't do, as it yields varying proportions without any consistent pattern.

(7) By simply imitating and generalizing the histogram structure hidden within the digital arrangement of Benford's Law — as shall be seen in the following chapters — the generic idea of the bin scheme was invented. What is needed here in the discovery of a consistent pattern is applying a set of infinite histograms with expanding bin width as opposed to having merely a single histogram; and instead of focusing on particular ranges according to any given data set; the universal range from zero to infinity shall become the focus for all the data sets. In essence, Benford's Law gave the clue for GLORQ. It was reverse engineering of sorts.

Chapter 59

The Base Invariance Principle in Benford's Law

Benford's Law is valid for all positional number systems of whatever base B; including of course our base 10 system. This important generalization is accomplished by using the appropriate logarithm base B in the expression for the probabilities for the first leading digit d (in the range of 1 to $B - 1$). The generalized Benford's Law for any base B is

Probability[First Leading Digit is d] $= LOG_B(1 + 1/d)$

As an example, for a positional number system with base 4, the three possible first digits are $\{1, 2, 3\}$. Here $LOG_4(1 + 1/1) = 0.50$; $LOG_4(1 + 1/2) = 0.29$; and $LOG_4(1 + 1/3) = 0.21$.

For a positional number system base 6 with $\{1, 2, 3, 4, 5\}$ as the set of all possible first digits, Benford's Law predicts the digital probabilities of $\{38.7\%, 22.6\%, 16.1\%, 12.5\%, 10.2\%\}$. This set is calculated as $\{LOG_6(1 + 1/1), LOG_6(1 + 1/2), LOG_6(1 + 1/3), LOG_6(1 + 1/4), LOG_6(1 + 1/5)\}$.

For base 2 number system, all numbers necessarily start with digit 1, hence the probability of digit 1 leading is 100%. This is also confirmed via the expression $LOG_2(1+1/1) = LOG_2(2) = 1$, namely 100%, and here Benford's Law is reduced to a tautology.

Chapter 60

The Scale Invariance Principle in Benford's Law

Remarkably, Benford's Law is valid assuming the use of any scale in measuring the physical phenomenon generating the data set under consideration. Surely, there is nothing special about kilograms, meters, seconds, hours, inches, or miles; they are all arbitrary; and philosophically, for the law to be considered universal and consistent, the law should hold true for any units and scales. The fact that Benford's Law is indeed independent of societal scale system renders it universality. This property of the law is called "The Scale Invariance Principle". Since Benford's Law shall be shown to be a special case and a consequence of GLORQ, the expectation is that its parent law GLORQ should also be scale invariant, and indeed GLORQ is found to be scale invariant as well. Surely, the small is beautiful phenomenon is totally independent of societal scale system.

As an example, the data on the time in the units of seconds between all successive earthquakes worldwide for the year 2012 shall be converted into units of minutes and into units of hours.

1st Digits of Earthquake Data in Seconds —
$\{29.9, 18.8, 13.5, 9.3, 7.5, 6.2, 5.8, 4.8, 4.2\}$

1st Digits of Earthquake Data in Minutes —
$\{28.6, 17.7, 12.8, 10.5, 8.6, 6.8, 5.8, 5.0, 4.2\}$

1st Digits of Earthquake Data in Hours —
$\{29.2, 16.9, 12.3, 9.8, 8.1, 7.0, 6.2, 5.7, 4.9\}$

The sign "%" is omitted here for brevity. The small deviations seen here for these three scales are due to the fact that the earthquake data set is not perfectly Benford and that it has finite number of data points. Had it been perfectly Benford and with infinite number of data points (or at least with a truly huge number of points) then there would be (almost) no deviations at all. It should be noted that a scale change from seconds to minutes say, entails multiplying each time interval quoted in seconds by the factor of 1/60. Hence, the more general view of the Scale Invariance Principle is that digit distribution of Benford-type data remains nearly unchanged under a multiplicative transformation of all the data points by the same multiplicand (i.e. the factor).

Chapter 61

Integral Powers of Ten (IPOT)

Integral Powers of Ten (IPOT) play a crucial role in the understanding of Benford's Law.

An integral power of ten is simply 10^{INTEGER} with either a negative or positive integer, as well as a zero power. For example, the IPOT numbers 0.001, 0.01, 0.1, 1, 10, 100 are directly derived from 10^{-3}, 10^{-2}, 10^{-1}, 10^{0}, 10^{1}, 10^{2}.

Adjacent integral powers of ten are a pair of two neighboring and consecutive IPOT numbers 10^{INTEGER} and $10^{\text{INTEGER}+1}$ such as 1 and 10, or 100 and 1000, and so forth.

Chapter 62

Benford's Law as a Special Case and Direct Consequence of GLORQ

With hindsight about GLORQ, which itself was obtained via the clues provided by Benford's Law, it shall now be demonstrated how the mathematical expression for GLORQ directly implies the mathematical expression of $LOG(1 + 1/d)$ of Benford's Law. But in order to accomplish that, it is necessary first to present the first digits distribution of Benford's Law as a particular bin scheme containing infinitely many expanding histograms.

Clearly, all numbers from **1** up to **2** such as 1.00, 1.15, 1.49, 1.76, 1.93, and 1.99 are with first digit 1; all numbers from **10** up to **20** such as 10.0, 13.8, 16.8, 18.2, and 19.6, are with first digit 1; and all numbers from **100** up to **200** such as 100, 141, 176, 195, and 198 are with first digit 1.

In general, the count of numbers within a given data set with first digit 1 is equivalent to the count of data points falling within $\ldots [0.01, 0.02), [0.1, 0.2), [1, 2), [10, 20), [100, 200) \ldots$ and so forth. The count of numbers within a given data set with first digit 7 is equivalent to the count of data points falling within $\ldots [0.07, 0.08), [0.7, 0.8), [7, 8), [70, 80), [700, 800) \ldots$ and so forth. Surely infinitely many other sub-intervals on the left and on the right of the above-mentioned five sub-intervals should also be included, such as $[0.001, 0.002)$ for say the particular number 0.00176, or $[7000, 8000)$ for say the particular number 7231, and so forth.

The brief numerical-to-digital exploration above then leads to portray Benford's Law as a particular 9-bin scheme with an inflation factor 10.

Indeed, Benford digital law can be reduced to a purely quantitative law when interpreted as the quantitative aggregation of an infinite set of histograms — each of nine bins — constructed between ...$0.001, 0.01, 0.1, 1, 10, 100, 1000, ...$ and so forth. The widths of these histograms are expanding by a factor of 10 with each consecutive histogram to the right.

Hence, first digits distribution of a given data set is nothing but the condensed or aggregated histogram of the infinitely many nine-bin local histograms standing between integral powers of ten.

For better and more concrete visualization of these generic and abstract histograms of Benford's Law, Figures 4.9–4.17 are included in Chapter 73.

This vista practically ignores the digits altogether! It focuses on the numerical data itself. Well, except for the fact that these mini histograms are deliberately constructed over a very particular partition of the entire x-axis range according to the cyclical way first digits occur.

Figure 4.6 depicts the first two cycles of an infinite nine-bin scheme starting from 1, with an inflation factor of 10, and an initial bin width of 1 unit. The bins for these two cycles are:

$$\{[1, 2), [2, 3), [3, 4), [4, 5), [5, 6), [6, 7), [7, 8), [8, 9), [9, 10)\}$$

$$\{[10, 20), [20, 30), [30, 40), [40, 50), [50, 60), [60, 70), [70, 80),$$
$$[80, 90), [90, 100)\}$$

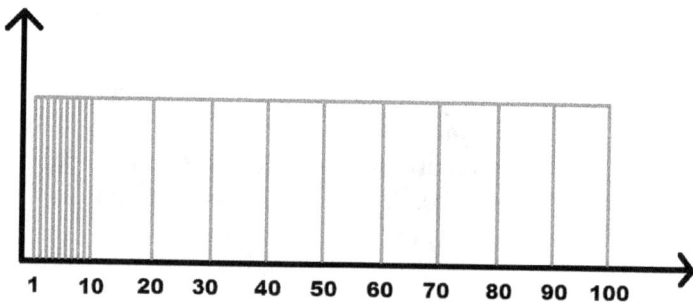

Figure 4.6:   Benford's Law as a Nine-Bin Scheme with an Inflation Factor of 10

A moment thought would convince anyone that the bin scheme of Figure 4.6 perfectly corresponds to first digits count, given that the scheme starts at 0 instead of starting at 1, having an infinitesimally tiny initial bin width, approaching the value of 0 in the limit.

The endpoints of the cycles here are $\ldots 0.001, 0.01, 0.1, 1, 10, 100, 1000, 10000, \ldots$ and so forth.

This implies an inflation factor of 10, as in the progressions $0.01 \times (\mathbf{10}) \to 0.1$, $0.1 \times (\mathbf{10}) \to 1$, $1 \times (\mathbf{10}) \to 10$, $10 \times (\mathbf{10}) \to 100$, $100 \times (\mathbf{10}) \to 1000$, and so forth.

In general, for any positional number system base B with $\{1, 2, 3, \ldots, (B-1)\}$ as the set of all possible first digits, the bin cycles expand by the inflation factor of B, and the number of bins in each cycle is $(B-1)$. For example, for our base 10 number system: $F = 10$ and $D = 9$. For base 4 number system: $F = 4$ and $D = 3$, and for base 16 number system: $F = 16$ and $D = 15$.

In conclusion: first digits proportions for any positional number system base B is equivalent to the proportions for bin schemes with $\mathbf{F} = (B)$ and $\mathbf{D} = (B - 1)$ or equivalently $B = D + 1$. Hence for number systems and their first digits, the corresponding bin scheme is such where D is always 1 less than F, or equivalently where F is always greater than D by 1, and consequently all such bin schemes tailor-made for positional number systems are constrained by $\mathbf{F} = \mathbf{D} + \mathbf{1}$.

With such succinct correspondence between these restricted bin schemes of the GLORQ and first digits distributions, it is now straightforward to demonstrate that digital Benford's Law is simply a special case and a consequence of the General Law of Relative Quantities when bin schemes are constructed under the constraint $F = D + 1$. The term F is then substituted by $D + 1$ everywhere in expression of GLORQ, yielding:

$$\text{GLORQ} = \frac{\ln\left(\dfrac{D + d(F - 1)}{D + (d - 1)(F - 1)}\right)}{\ln(F)}$$

$$= \frac{\ln\left(\dfrac{D + d(D + 1 - 1)}{D + (d - 1)(D + 1 - 1)}\right)}{\ln(D + 1)} = \frac{\ln\left(\dfrac{D + d(D)}{D + (d - 1)(D)}\right)}{\ln(D + 1)}$$

$$= \frac{\ln\left(\dfrac{D(1 + d)}{D(1 + (d - 1))}\right)}{\ln(D + 1)} = \frac{\ln\left(\dfrac{1 + d}{1 + (d - 1)}\right)}{\ln(D + 1)} = \frac{\ln\left(\dfrac{1 + d}{d}\right)}{\ln(D + 1)}$$

$$= \frac{\ln\left(\dfrac{1}{d} + \dfrac{d}{d}\right)}{\ln(D + 1)} = \frac{\text{LOG}_e\left(1 + \dfrac{1}{d}\right)}{\text{LOG}_e(\text{ BASE })} = \frac{\text{LOG}_{\text{BASE}}\left(1 + \dfrac{1}{d}\right)}{\text{LOG}_{\text{BASE}}(\text{BASE})}$$

$$= \frac{\text{LOG}_{\text{BASE}}\left(1 + \dfrac{1}{d}\right)}{1} = \text{Benford's Law}$$

It should be noted that $\ln(X)$ is the abbreviated version of $\text{LOG}_e(X)$. In addition, the logarithmic identity $\text{LOG}_A X = \text{LOG}_B X / \text{LOG}_B A$ is applied twice to convert the ratio $\text{LOG}_e / \text{LOG}_e$ of the natural logarithm base e, into the ratio $\text{LOG}_{\text{BASE}} / \text{LOG}_{\text{BASE}}$ of any other logarithmic base.

In addition, the logarithmic identity $\text{LOG}_B B = 1$ is used in the above derivation, as well as the relation $B = D + 1$ or equivalently $\text{BASE} = D + 1$.

For our positional number system base 10, the common logarithm base 10 is used to arrive at the expression $\text{LOG}_{10}(1 + 1/d)$ or simply as in the abbreviated form of $\text{LOG}(1 + 1/d)$.

Hence the phenomenon of Benford's Law is shown to be physical and quantitative, existing independently of our artificially invented positional number system. Benford's Law is simply a sideshow to this physical law of nature (GLORQ) which can be measured and detected by ways other than our own digital perceptions. We are no longer seduced and blinded by the incredible efficiency of our number system, completed early in the Renaissance Period about 700 years ago. We are now able to acknowledge its arbitrariness, and do not err in believing that the ubiquitous and almost consistent distribution of its symbolic (first) digits as in $LOG(1+1/d)$ constitutes the ultimate say in data patterns. The ultimate data pattern is GLORQ!

Our positional number system is surely the most efficient scheme of counting and calculating quantities; and its perceived perfection can be succinctly expressed via our shared belief at this current epoch that another superior system which would ease calculation even further, or which would perform better in some new and unsuspected numerical and quantitative aspects, would never be discovered — that our civilization had arrived at the ultimate number system.

A demonstration of the incredible efficiency and perfection of our number system can be gotten by pointing out to the simple fact that this numerical system hasn't been revised at all in many centuries; it's still the same as it was in the late Renaissance Period; all the while our modes of cultivation, production, communication, and transportation, have been radically changed and improved. Perhaps it is an indication that there is simply nothing to improve with regard to our positional number system.

Our positional number system is structured in such a way that first digits cycles perfectly correspond to the particular bin scheme of GLORQ with $F = D + 1$. One might characterize this correspondence as a coincidence, and certainly that digit 0 inventor, and the others who contributed ideas and paved the way towards a fully functioning positional number system, never attempted or intended to correspond to any bin scheme or to discover any consistent statistical pattern in real-life data. Their goal was focused merely on creating an efficient number system, to ease counting and calculations. Yet, it so happened that the number system that we have invented turned out to have that structure of the bin schemes of GLORQ. Other number systems such as Roman numerals simply do not have the GLORQ bin structure for their first numeral cycles, and this is exactly why no statistical pattern could ever be found regarding their symbols and numerals when data is converted into these arcane number systems.

The number of bins D in GLORQ is equivalent to the number of all possible first digits in any positional number system, yet the letter D does not signify "Digit" as in the initial letter of the word, instead it signifies the more flexible and generic concept of the number of bins in the scheme. The crucial generalization of GLORQ over and above Benford's Law is that the restriction $F = D + 1$ of positional number systems is relaxed in order to include also the infinitely numerous possibilities of $F \neq D+1$ bin schemes, and those possibilities cannot be interpreted in any way as positional number systems and first digits. This is why the adjective "General" is included in the coining of the term "The General Law of Relative Quantities".

On the face of it, Benford's Law ignores quantities and obsessively focuses on symbolic digits. For example, 18133 is bigger than 526, yet the law considers 18133 as belonging to the lower class of first digit 1, and it considers 526 as belonging to the higher class of first digit 5. Nevertheless, ultimately digits signify quantities, and therefore the small is beautiful phenomenon, GLORQ, and Benford's Law, all integrate and accommodate each other.

First digit 7 indeed has bigger quantitative significance than first digit 4 when both reside within the same range having adjacent integral powers of ten as its edges, such as within (1, 10), or within (10, 100), and so forth. Within (10, 100), the number 73 belongs to a higher first digit class and it signifies a bigger quantity, as compared with the number 48 which belongs to a lower first digit class and which signifies a smaller quantity. In other words, for values residing exclusively within the range (10, 100) say, first digits indeed signify quantities, and the first digit rank perfectly corresponds to the quantitative rank!

Chapter 63

Benford and GLORQ as a Subset of the Small Is Beautiful Phenomenon

All the versions and variations of the processes leading to the small is beautiful phenomenon discussed in Section 2 are typically cumulative in nature. As a norm, they do not swiftly revolutionize quantities in favor of the small in one fell swoop from an original even and uniform quantitative configuration where all sizes are equally distributed. In other words, under the influence of these processes, quantitative configuration is gradually changing, slowly evolving, becoming skewer step by step, and lending the small an ever increasing portion of overall distribution. Typically but not always, saturation finally sets in, and quantitative configuration converges to some steady state (often GLORQ) beyond which no further quantitative changes are made. Chapter 52 discusses in detail saturations in GLORQ-inducing processes, and how extreme skewness is almost always precluded.

The generic causes of the small is beautiful phenomenon are: partition processes; multiplication processes; data aggregation; and chains of distributions. Assuming without any loss of generality that the initial data set upon which the generic processes are acting upon is of the uniform quantitative configuration, then in extreme generality, there are five possible scenarios:

(1) Plenty applications of the process but with limited final effect on quantities as early saturation leads to a mild manifestation

of the small is beautiful phenomenon, stopping well short of the skewer Benford and GLORQ configuration.

(2) Plenty applications of the process with far-reaching saturation point leading all the way exactly to the Benford and GLORQ configuration.

(3) Plenty applications of the process with far-reaching saturation point leading to a highly skewed quantitative configuration which is even skewer than that of the Benford and GLORQ configuration.

(4) Only very few applications of the process, stopping well short of saturation point itself, resulting in limited effect on quantities with only mild manifestation of the small is beautiful phenomenon, stopping well short of the skewer Benford and GLORQ configuration.

(5) Only very few applications of the quantitatively powerful process leading almost immediately and swiftly to some highly skewed configuration or to the Benford and GLORQ configuration.

It is noted that in all of the five scenarios above, the small is always more numerous to some extent, and sometimes it is even extremely numerous. Hence the small is beautiful phenomenon has by far much wider scope and it is much more prevalent in the physical world and in the realm of abstract mathematics than the more particular Benford and GLORQ quantitative configuration. This statement does not imply that Benford's Law and GLORQ are not prevalent in scientific, physical, and numerous other data types, on the contrary, they are highly prevalent. The statement only implies that in almost all the counter examples and exceptions to Benford's Law and GLORQ, the small is beautiful phenomenon is still valid, albeit with different quantitative configurations (which are typically milder, but at times even skewer).

The assertion about these five highly abstract and generic scenarios above is derived from more concrete experience with numerical examples and from general research in Benford's Law. Some of the above scenarios have already been discussed in earlier sections. While this discussion may sound vague, in fact it is rather a very essential overview of the entire quantitative phenomenon of this book. For those statisticians and data analysts who have worked on data sets and the Benford phenomenon for many years, including doing theoretical research, this generic statement seems natural, fundamental, and quite necessary.

All this can be stated more succinctly in three ways:

  (I) The Benford's Law and GLORQ configuration is a subset of the small is beautiful phenomena.

 (II) The small is beautiful phenomenon is even more prevalent than Benford's Law and GLORQ.

(III) A significant portion of non-Benford and non-GLORQ real-life data is quantitatively structured in the spirit of the small is beautiful phenomenon.

Chapter 64

Physical Order of Magnitude
of Data (POM)

Rules regarding expectations of compliance with Benford's Law and GLORQ rely heavily on measures of order of magnitude and variability of data, therefore this chapter and the next one shall be devoted to these prerequisites and essential topics.

Physical order of magnitude of a given data set is a measure that expresses the extent of its variability. It is defined as the ratio of the maximum value to the minimum value. The data set is assumed to contain only positive numbers greater than zero.

Physical Order of Magnitude (POM) = Maximum/Minimum

The classic definition of order of magnitude involves also the application of the logarithm to the ratio maximum/minimum, transforming it into a smaller and more manageable number.

Order of Magnitude (OOM) = LOG_{10}(Maximum/Minimum)

Order of Magnitude (OOM) = LOG_{10}(Maximum) − LOG_{10}(Minimum)

Since such logarithmic transformation has a monotonic one-to-one relationship with max/min, it does not provide for any new insight or information, but could rather be looked upon in a sense as simply the use of an alternative scale, still measuring the same thing. For this reason, the complexity of logarithm can be avoided altogether by referring only to the simple POM measure.

The more profound reason for using POM instead of OOM is its feature as a universal measure of variability, totally independent of societal number system in use, as well as being independent on the arbitrary choice of base 10, derived from the chanced or random occurrence of us having 10 fingers. This is the motivation behind the use of the term "physical", expressing real and physical measure of variability, divorced from any numerical inventions, and especially so when data relates to the natural world such as in scientific figures and physical information.

OOM is perhaps more appropriate for a single isolated number, where it is re-defined as simply LOG_{10}(Number) without any reference to maximum, minimum, or any ratio. If we can assume that that number is an integral whole number without any [trailing] fractional part, then an alternative meaning of this OOM definition is simply expressing how many digits approximately are necessary to write the number. Surely in general, the bigger the [integral] number the more digits it takes to write it! For example, $LOG_{10}(8,200,135) = 6.9$, which is about 7, and that's exactly how many digits the number involves. As another example, $LOG_{10}(10,000,000) = 7.0$ which is exactly one digit less than the number of digits involved in writing the number, namely 8 digits.

Let us return the focus to data sets, containing many numbers, a minimum, and a maximum.

Suppose that an extensive database on the size (height) of a certain **giraffe species** somewhere on the African Savanna has a minimum of 5.0 meters; a maximum of 6.0 meters; and an average of 5.5 meters. Here the range of variability for the giraffe is $(6.0 - 5.0) = 1.0$ meters.

Suppose that an extensive database on the size (length) of a particular **ant species** somewhere in the Amazon Basin has a minimum of 0.003 meters; a maximum of 0.009 meters; and an average of 0.006 meters. Here the range of variability for the ant is $(0.009 - 0.003) = 0.006$ meters.

Which species should be considered as more varied in size, the ant or the giraffe? The giraffes vary over 1 meter, while the ants vary over only 0.006 meters, and Standard Deviation of the giraffes is much bigger than Standard Deviation of the ants, therefore on the face of it, the giraffe has more variability than does the ant. Yet, such an absolute approach, imposing the same universal benchmark of the scale of the meter on the two very different species is quite arbitrary.

What is necessary here is to measure <u>auto-variability</u>, namely the variability of the data relative to itself, not relative to any arbitrary measure of the meter or other artificial scales, nor a comparison relative to other data sets regarding other species.

That apparent "huge" variability of 1.0 meter for the giraffe is actually quite small in comparison with its own average of 5.5 meters, namely merely $(1.0)/(5.5)$ or 18%.

That apparent "tiny" variability of only 0.006 meter for the ant is actually quite huge in comparison with its own average of 0.006 meters, namely a whopping $(0.006)/(0.006)$ or 100%.

The POM expression as the ratio of maximum to minimum also represents an auto-variability measure regarding the data set itself, and it is independent of any other data sets regarding different organisms and entities, as well as independent of any societal number system or base in use. POM is certainly also unit-less, independent of any societal and arbitrary units and scales, such as the meter, centimeter, kilometer, inch, or mile.

For the giraffes, POM measure is 6/5 or merely 1.2. For the ants, POM measure is a 0.009/0.003 or 3.0. Therefore, POM measure of the ants is approximately three times larger than POM measure of the giraffes!

In the extreme case where all the numbers in the data set are identical, having the value R say, variability is then nonexistent, and POM = maximum/minimum = $R/R = 1$.

Chapter 65

A Robust Measure of Physical Order of Magnitude (CPOM)

It is perhaps unfortunate that the literature in statistics does not seem to contain any robust definition of order of magnitude. Such a measure should prove steady and consistent for all types of data sets, strongly resisting outliers, preventing them from overly influencing the numerical measure of data variability.

In order to accomplish exactly that, and also to preserve the advantage of avoiding dependencies on arbitrary societal number systems and particular bases, the basic (independent) structure of POM shall be used, but with the added modification of simply eliminating any possible outliers on the left for small values and on the right for big values. This is accomplished by narrowing the focus exclusively onto the core 80% part of the data. This brutal purge eliminates without mercy any malicious and misleading outliers as well as any innocent and proper data points which happened to stray just a little bit away from the core part of the data. The measure shall be called Core Physical Order of Magnitude and it is defined as follows:

Core Physical Order of Magnitude (CPOM) = $P_{90\%}/P_{10\%}$

The definition simply reformulates POM by substituting the 10th percentile (symbolized as $P_{10\%}$) for the minimum, and by substituting the 90th percentile (symbolized as $P_{90\%}$) for the maximum.

The 10th percentile is the value below which about 10% of the data points may be found. The 50th percentile is the median, below which about half of all the data points may be found. The 90th percentile is the value below which about 90% of the data points may be found.

As an example, 50 data points, sorted low to high, shall be examined, as shown below:

2	23	24	25	26	27	28	29	32	33
33	33	34	36	37	38	38	39	40	41
42	47	48	50	51	52	53	55	56	57
59	60	63	67	68	75	76	77	78	79
80	84	86	91	94	103	107	114	**213**	**567**

There are three obvious outliers within this data set, namely $\{2, 213, 567\}$ shown above in bold font for emphasis. Calculating POM mindlessly without any worries whatsoever about possible distortions from outliers, leads to $(567)/(2) = 283.5$. But this value greatly exaggerates the variability of the data set which spans mostly the much narrower range of 23 to 114 in the approximate. The 10th percentile here is 26.9 and the 90th percentile is 94.9. It follows that $\text{CPOM} = P_{90\%}/P_{10\%} = (94.9)/(26.9) = 3.5$. This is by far a more realistic value for the true variability of the data set, focusing on the core 80% of the data.

Figure 4.7 depicts the histogram of the above data set, showing the 3 outliers as short, thick, and black lines, as well as the 10th and 90th percentile points utilized in the definition of CPOM.

An alternative measure for the core variability of any given data set could be defined without involving any percentiles at all. This measure would simply eliminate first all outliers before any calculation of the originally defined POM measure is performed, and then applying the newly observed minimum and maximum values after the elimination of outliers took place.

Hence for the above data set, once $\{2, 213, 567\}$ are eliminated, POM for the rest of the data is simply $(114)/(23) = 4.9$. For this data set it is perhaps straightforward to classify $\{2, 213, 567\}$ as outliers, and almost all data analysts should strongly agree with this classification. Yet, decisions and considerations regarding which data points constitute outliers and which belong to the core data for this or that particular data set could be considered at times as subjective

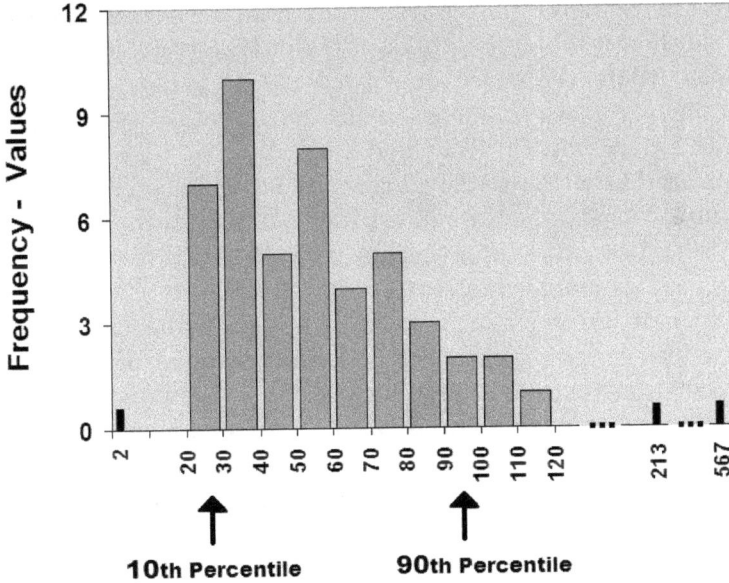

Figure 4.7: Core Variability Focusing on the Central 80% of Data

and personal, depending on the data cases. Hence the motivation for having a universal and objective guideline here; a benchmark which would encompass all data sets equally, and which is provided by CPOM.

By strictly applying the 10th and 90th percentiles for all data sets, the procedure avoids vague definitions and arguments about outliers. Surely, the price paid for such universality in the procedure is the occasional eliminations of innocent data points near the edges which are actually very much part of the data, but are being swept away by the crude cleansing method of CPOM.

This is akin to a malignant cancer surgery where the surgeon is keen on making sure that no cancer cells whatsoever are left in the area, so that remission is to be avoided at all cost, and therefore the surgeon is cutting some more all around the tumor, even in the healthy tissue and cells immediately surrounding it. For the data set above, the crude outlier-surgery of CPOM removes also the innocent points $\{23, 24, 25, 26, 103, 107, 114\}$ which are actually authentic part of the data, but this is the price we are willing to pay in order to standardize the procedure.

For a more liberal procedure, with stronger emphasis on avoiding losing authentic data points, and less emphasis on eliminating all possible outliers, the wider range from the 5th percentile to the 95th percentile might be considered, applying the ratio $P_{95\%}/P_{5\%}$ as the measure for the core 90% of the data.

More recently, the author has switched to an even more liberal procedure, focusing on the much wider range from the 1st percentile to the 99th percentile, applying the ratio $P_{99\%}/P_{1\%}$ as the measure for the core 98% of the data.

In general, the rejection of outliers appearing in a given data set may be justified, or it may actually be misguided. For example, if over 50,000 students at a large university are surveyed with regards to height, and the top value is say 7.25 meters, then this outlier is certainly some kind of an error in recording and should be excluded from further analysis. If the top value is say 2.37 meters, then this 2.37 outlier is actually an integral part of the data set, and especially so if the well-known tall student is ordered to appear at the administration office, rudely interrupting his exciting basketball game at the court, and another measurement is taken, confirming his 2.37 meter height as well as his existence. This is not simply a matter of mere semantics, and there is a compelling argument not to classify this 2.37 value as an outlier, although in reality it depends on the context.

Chapter 66

Two Essential Requirements for Benford and GLORQ Behavior

One of the two essential prerequisites or conditions for data configuration with regards to compliance with Benford's Law and GLORQ is high variability and that the value of the order of magnitude of the data set should be approximately over 3; in other words, that LOG_{10}(Maximum/Minimum) > 3, and that therefore (Maximum/Minimum) $> 10^3$. This in turn implies that the threshold POM value (separating compliance from non-compliance) is about 1000, namely that the prerequisite of POM > 1000 constitutes the condition for compliance.

The above prerequisite for compliance totally ignores the thorny issue of outliers and edges, and in that sense it is too simplistic and even completely erroneous for some data sets. Hence, using the CPOM qualification is essential in judging whether or not a given data set is expected or not expected to comply with Benford's Law. The proper qualification for expectance of compliance with the law in the approximate — obtained via extensive empirical studies — is then as follows:

Core Physical Order of Magnitude = $P_{90\%}/P_{10\%} > 100$.

Actually, even lower CPOM values such as 50 and 30 are expected to yield Benford, but falling below 30 does not bode well for getting anywhere near the Benford configuration.

Skewness of data where the histogram comes with a prominent tail falling to the right is the second essential criterion necessary for Benford and GLORD behavior. Indeed, most real-life physical data sets are generally skewed in the aggregate, so that overall their histograms have tails falling on the right, and consequently the quantitative configuration is such that the small is numerous and the big is rare, while low first digits decisively outnumber high first digits.

The *asymmetrical*, Exponential, Lognormal, k/x are typical examples of such quantitatively skewed configuration, and therefore they are approximately, nearly, or exactly Benford — respectively. The *symmetrical* Uniform, Normal, Triangular, Circular-like, and other such distributions are inherently non-Benford, or rather anti-Benford, as they lack skewness and do not exhibit any bias or preference towards the small and the low.

Symmetrical distributions are always non-Benford, no matter what values are assigned to their parameters. By definition they lack that asymmetrical tail falling to the right, and such lack of skewness precludes Benford behavior regardless of the value of their order of magnitude. Order of magnitude simply does not play any role whatsoever in Benford behavior for symmetrical distributions. For example, first digits of the Normal(10^{35}, 10^8) or the Uniform(1, 10^{27}) are not Benford at all, and this is so in spite of their extremely large orders of magnitude. In summary: Benford behavior in extreme generality can be found with the confluence of sufficiently large order of magnitude together with skewness of data — having a histogram falling to the right. The combination of skewness and large order of magnitude is not a guarantee of Benford behavior, but it is a strong indication of likely Benford behavior under the right conditions. Moderate [overall] quantitative skewness with a tail falling too gently to the right implies that digits are not as skewed as in the Benford configuration. Extreme [overall] quantitative skewness with a tail falling sharply to the right implies that digits are severely skewed, even more so than they are in the Benford configuration. The only one exception to the generic rule above requiring skewness as well as high order of magnitude for Benford behavior is the perfectly Benford k/x distribution defined over adjacent integral powers of 10 such as (1, 10) having the very low order of magnitude value of 1, as shall be discussed in the next chapters.

Bowley Skewness for example, defined as $[(Q_3 - Q_2) - (Q_2 - Q_1)]/[Q_3 - Q_1]$ is an intuitive measure of skewness but its numerical value fluctuates greatly across data sets. Calculated Bowley Skewness values for numerous Benford data sets and distributions do not yield any consistent result, except that all values come out above 0.3 and below 1.0, and which is consistent with the fact that all Benford data sets are positively skewed in the aggregate.

In contrast to Bowley's unstable value for Benford data sets, Benford's Law and GLORQ are very consistent and almost exact measures of skewness, with very little fluctuations across Benford data sets.

Chapter 67

Sum of Squared Deviation Measure (SSD)

It is necessary to establish a standard measure of "distance" from the Benford digital configuration for any given data set. Such a numerical measure could perhaps tell us about the conformance or divergence from the Benford digital configuration of the data set under consideration. This is accomplished with what is called "Sum of Squared Deviations" (SSD) defined as the sum of the squares of the "errors" between the Benford expectations and the actual/observed values (in percent format — as opposed to fractional/proportional format):

$$\text{SSD} = \sum(\text{observed \% of digit } d - 100 \times \text{LOG}(1 + 1/d))^2$$

with d running from 1 to 9. For example, for the observed first digits proportions as in $\{31.1, 18.2, 13.3, 9.4, 7.2, 6.3, 5.9, 4.5, 4.1\}$, SSD measure of distance from Benford is calculated as

$$\textbf{SSD} = (31.1 - \textbf{30.1})^2 + (18.2 - \textbf{17.6})^2 + (13.3 - \textbf{12.5})^2$$
$$+ (9.4 - \textbf{9.7})^2 + (7.2 - \textbf{7.9})^2 + (6.3 - \textbf{6.7})^2 + (5.9 - \textbf{5.8})^2$$
$$+ (4.5 - \textbf{5.1})^2 + (4.1 - \textbf{4.6})^2 = \textbf{3.4}$$

SSD generally should be below 25; a data set with SSD over 100 is considered to deviate too much from Benford; and a reading below 2 is considered to be ideally Benford.

Applying the SSD measure to five data sets from Chapter 45, namely Population, Price List, Exponential Growth, Lognormal, and Earthquakes, yields the low SSD values of 1.3, 5.8, 2.3, 1.9, and 3.1 respectively. For each of these five data sets the value of SSD is below 6, indicating strong compliance with Benford's Law.

This SSD measure can be easily generalized for higher order digit distributions, for other bases in positional number systems, and for empirical GLORQ results of whatever F and D values, as follows:

$$\text{SSD} = \sum(\text{Observed\%} - \text{Theoretical\%})^2$$

with the summation index running from 0 to 9 for higher digital orders in our decimal number system. For the first order digits in number systems with bases other than 10, the summation index runs from 1 to (Base − 1). For GLORQ, the summation index runs from 1 to the number of bins, namely D.

Chapter 68

A Critique on the Mixture of Distributions Model in Benford's Law

The two discoverers of the law also attempted to give explanations for this digital phenomenon, one involving random divisions of two Uniform Distributions, and the other involving aggregating particular data structures. Apparently, the mathematical community chose to ignore or reject these two explanations as irrelevant models, not corresponding to the structure of typical real-life data, all the while greatly appreciating the digital discovery itself as relevant and certainly worthy of study. Subsequently, an explanation assuming the existence of some scale-invariant digital law was also rejected on the grounds that such an assumption cannot be proven. Another attempt at an explanation is the mathematical demonstration that a large mixture of distributions, each defined over the positive x-axis, obeys Benford's Law in the limit as the number of distributions goes to infinity. This is also called "the distribution of all distributions".

First, a large collection of random distributions is assembled randomly, such as Uniform(0, 5), Normal(17, 4), Uniform(3, 16), Exponential(0.02), Lognormal(8, 1), Lognormal(5, 6), Exponential(1.9), and so forth. Then a repetitive process generating random numbers in stages starts in earnest. In each stage, one particular distribution is randomly selected [as if uniformly and evenly distributed], and then one random realization from that selected distribution is simulated and collected. This process continues until numerous such random values have been collected.

Translating this abstract model into real-life data collection reveals the irrelevance of the whole idea. Unfortunately the model can only be applied to data blindly and randomly collected from a large variety of sources, and only when very few numbers are picked from each source. The sources should consist of positive values exclusively. Strictly speaking the process should pick only one number from a given source, then another number from another (related or totally unrelated) source, and so forth. The end result of this whole process is a large mixture of unrelated numbers, representing a meaningless data set not conveying any specific message or information, yet digitally structured as Benford. The model cannot be a valid explanation though for the almost perfectly Benford data sets regarding single-issue (*single-source, non-mixed*) physical and scientific phenomena, such as the time between earthquake occurrences, or river flow, or population count, or pulsar rotation frequency, or half-life of radioactive material, and so on. The mixture of distributions model does not show any immediate or obvious relation to the Benford way Mother Nature generates her physical quantities. It is very hard or rather impossible to argue that river flow, earthquake timing, or pulsar rotation, are the results of some aggregation of numerous invisible, mysterious, and unrelated, mini distributions.

On the other hand, the explanation of the phenomenon provided by the model regarding multiplication processes is not only plausible in general, but it is undoubtedly the proper model in many real-life cases. Isaac Newton's multiplicative expression of $F = M \times A$, among many other such expressions for gravitational and electromagnetic forces, including numerous applications in physics, chemistry, other sciences, and engineering involving multiplicands, all remind us of the direct connection between the multiplicative model and real-life physical data. Partition model as the explanation of the Benford's Law phenomenon in many real-life cases is also certainly correct and plausible. This is especially so in light of the occasional interpretations of "partition" as "composition" or "consolidation", and which endows partition model by far greater scope of applicability. In the same vein, data aggregation and the

MEASURE	VALUE	UNIT
Time between 2 earthquakes - 1/27/2013 - 7:05 & 7:37	31.9	Second
Depth below the ground - Earthquake - 12/30/2013 - 23:87	5.3	Kilometer
Rotation frequency of pulsar IGR J17498-2921	401.8	hertz
Amazon river length	6,437	Kilometer
New York City Population 2013 Census	8,337,342	People
Temperature of the star Polaris	6,015	Kelvin
Mass of exoplanet Tau Ceti f Constellation Cetus	0.783	Solar mass

Figure 4.8: Mixture of Real-Life Data Sets Mimicking the Mixture of Distributions Model

related chain of distributions model are surely the obvious and plausible models for numerous real-life cases.

In order to demonstrate very clearly the lack of statistical meaning for data derived in the spirit of the mixture of distributions model, a concrete example is given, with only the first 7 numbers of that "infinite" or sufficiently large collection of values shown in Figure 4.8.

The set of pure values with only the first 7 numbers shown is then:

$$\{31.9, \ 5.3, \ 401.8, \ 6437, \ 8337342, \ 6015, \ 0.783, \ ...\}$$

But surely this set does not represent seconds, hertz, solar mass, or any other quantity, nor does it convey any specific data-related message. This set does not represent any particular physical entity or physical concept, nor does it stand for any single scale. Mathematically though, this set is demonstrated to be perfectly Benford in the limit, yet, it explains nothing but itself.

The author has spent the better part of a week in February 2012 randomly gathering numbers from all sorts of websites, in the first ever serious attempt at empirical verification of this model. In total 34,269 positive numbers were obtained. The results confirmed the theoretical expectation, with first digit distribution coming at $\{28.8, 16.4, 12.4, 9.8, 8.3, 7.3, 6.1, 5.7, 5.3\}$, and with SSD of the low value of 4.6, indicating strong compliance with Benford's Law.

Detailed discussions about this data experiment, the mathematician who in 1995 provided the proof of the mixture of distributions (*Proceedings of the American Mathematical Society* 123, *The American Mathematical Monthly* 102, *Statistical Science* 10(4)), and the mixture of distributions harmonious connection to other issues in Benford's Law, could be found in Chapters 55, 69, 110, of the author's 2014 book on Benford.

A critique on this empirical study might claim that the result thus obtained actually confirms the principle behind data aggregations (such as house number in address data), as opposed to confirming the validity of the mixture of distributions model. The essential distinction between data aggregations and mixture of distributions is the point where each data set or each distribution starts. For data aggregations the start is at 0 or 1, or perhaps at another very low value. For the mixture of distributions the starts could be at any value, big or small, and without any restriction. It is possible to argue that most of the relevant websites contain values that typically start at very low values that are not much higher than 0 or 1, and that therefore the empirical study pertains to the model of data aggregations, rather than to the mixture of distributions model.

Chapter 69

The Random and the Deterministic Flavors in Benford's Law

Not all Benford data sets are created equal, but rather they come with two distinct flavors, the random flavor and the deterministic flavor. The essential distinction between these two flavors is the way digits behave throughout the entire range of the data locally, on smaller sub-intervals.

If a given data set with a range say between 3 and 2789 is nearly perfectly Benford, could we then conclude that small parts of the data are also Benford, just because they have been cut out from a whole Benford configuration? For example, is the sub-set of the data belonging to the sub-range between 3 and 1652 also Benford? Is the sub-set of the data belonging to the sub-range between 10 and 100 also Benford? Does the whole endow its Benford property to its parts?

Globally, from the minimum on the very left part of the range, all the way to the maximum on the very right part of the range, digits proportions are as in $LOG(1 + 1/d)$ overall, as predicated by Benford's Law. Yet, for random-flavored data, local digits distributions on smaller sub-intervals show a remarkably consistent pattern of differentiation, as digits develop from near digital equality on the left for low values, to approximately the Benford digital configuration around the middle, and finally to extreme digital inequality on the far right for big values, where low digits overwhelm high digits, and where digit 1 typically usurps leadership by earning

40% or even more than 50% proportion in some cases. This pattern in random-flavored data is coined as "Digital Development Pattern".

In order to be able to observe Digital Development Pattern, it is necessary to partition the relevant section of the x-axis into sub-intervals standing between integral powers of ten, such as 0.01, 0.1, 1, 10, 100, and so forth. This partition is the most natural one in the context of Benford's Law since these points signify the beginnings and the ends of all the first digit cycles.

The vast majority of real-life data is of the random flavor. A tiny minority — such as deterministic exponential growth series and data relating to k/x distribution — come with very consistent local digit distributions throughout the entire range of data, namely that of the Benford digital configuration, which is found equally on the left, in the center, and on the right, without any development or changes whatsoever.

The coining of the terms "deterministic" and "random" is usually appropriate in most cases, but these terms should not be taken literally, because the distinction here is actually not about randomness in data versus predictable events and deterministic generation of resultant numbers, but rather about localized digital behavior within the entire data range. The choice of these two terms is due to the fact that almost all random data come with such differentiated local digital behavior, while the particular case of deterministic and predictable exponential growth series comes with the consistent Benford behavior throughout its entire range. But these two terms would seem awkward when random data has (very rarely) the consistent Benford behavior throughout its entire range [*such as in the case of exponential growth series with variable and random growth rate, as well as in the case of k/x statistical distribution*], or when deterministic data comes with digital development. Perhaps future authors would coin the alternative terms of the "consistent flavor" and the "developmental flavor" in Benford's Law.

Uniformity of mantissa is often referred to as the "General Form of Benford's Law". In the most simplistic way, mantissa could be described as "the fractional part of the logarithm", although this definition is not true for numbers less than 1 having negative log values. The formal definition of the mantissa of any positive number X is that unique solution to the relation $X = 10^C \times 10^{\text{MANTISSA}}$,

C being an integer called the "characteristic" which is obtained by rounding down $LOG(X)$ to the nearest integer, namely the largest integer less than or equal to $LOG(X)$. Mantissa has a one-to-one correspondence with digital configuration, all orders considered. Therefore, instead of stating in great details how all the digital orders are distributed, one might as well consider more concisely how mantissa is distributed, and as a consequence all digital orders are determined in one fell swoop. As discussed in Kossovsky (2014) Chapters 61, 62, and 63, the main result regarding mantissa is: Benford's Law implies uniformity of mantissa and uniformity of mantissa implies Benford's Law. Hence, data with uniform and flat histogram of its logarithm can be perfectly Benford assuming properly defined ranges, as it can easily yield uniformity of mantissa. Data with rising log histogram is approximately with digital equality, while data with falling log histogram has severe digital inequality in favor of low digits, more so than the inequality of the Benford digital configuration.

The "random flavor" is one where the histogram of the logarithm of the data curves around, starting from [or very near] the log-axis itself, then rising, then reaching a brief plateau at its maximum value, then falling, and finally terminating at [or very near] the log-axis again further to the right. This meticulous upside-down-U behavior of the histogram of the logarithm of data is nearly a universal feature in practically all random data sets, being even more ubiquitous than Benford's Law itself! The rising portion of the log histogram on the left side yields an approximate digital equality locally, and the falling portion of the log histogram on the right side yields extreme digital skewness locally in favor of low digits, much more so than in the Benford configuration, both sides canceling and offsetting each other, hence the empirically observed differentiated digital behavior locally for random data sets, namely the Digital Development Pattern.

The "deterministic flavor" is one where the histogram of the logarithm of the data is flat and uniform. A data set which consistently obeys Benford's Law throughout all its parts and sub-intervals, namely data lacking the Digital Development Pattern, is characterized by having a flat and horizontal histogram for its logarithmically transformed data points. This is the case of exponential growth series and k/x distribution.

As it happened, log values of an exponential growth series are uniformly spread along the log-axis, albeit in a discrete fashion, and where distances between log values are constant. This fact clearly explains the absence of the Digital Development Pattern in exponential growth series.

It should be noted that the Digital Development Pattern is extremely prevalent in real-life data, even more so than the Benford phenomenon itself! In other words, practically all random data, Benford as well as non-Benford types [such as those with low order of magnitude say] clearly exhibit the Digital Development Pattern throughout their ranges. There is practically no exception!

The k/x distribution is the only density that perfectly obeys Benford's Law for a range standing between two adjacent IPOT points, such as (1, 10), (10, 100), or (100, 1000), and so forth. For such adjacent IPOT ranges, there exists no other distribution that perfectly obeys Benford's Law (with all higher orders considered) except k/x distribution! On such particular intervals the k/x distribution is unique!

It should be noted that k/x is also perfectly Benford whenever it is defined between any two points A and B such that log difference LOG(B) − LOG(A) is an integer greater than 1, such as say the interval (1.22835, 12283.5) where log difference is the integral value of 4, but k/x is not unique on such wider interval, and there are in principle infinitely many other distributions that are perfectly Benford as well.

Certainly the case of k/x distribution is quite exceptional in the field of Benford's Law, yet its highly-consistent Benfordian feature (totally lacking the Digital Development Pattern) renders it quite irrelevant to practically all types of real-life random data! "Paradoxically", k/x has loyally served us in deciphering GLORQ, which is the parent law of Benford!

Such is the seductive power of k/x distribution in the context of Benford's Law that some misguided authors and overly enthusiastic students of Benford's Law start their article or essay by basing it on some assumption or feature regarding the k/x distribution and then proceed to draw far reaching conclusions, mistakenly extrapolating the odd case of k/x to all real-life random data. Such regrettable trend has led to several erroneous conclusions, published in respectable journals, and officially certified by expert mathematicians as true. This author has taken on the dissenting role of an agitator as well as a prophet of doom, preaching the virtue of separating the random from the deterministic and of becoming aware of this crucial distinction in the field, and predicting the encountering of contradictions between the empirical and the theoretical in all such misguided pseudo-mathematical endeavors.

More generally for GLORQ in the context of development:

For "random data", bin developmental pattern is found regardless of the relative values of F and D variables. Yet, for bin schemes with F value much greater than D value, development is quite fast and very dramatic, while for bin schemes with F value less than D value, development is quite mild and slow.

For "deterministic data" such as the k/x distribution defined over an extremely large range or exponential growth series with an enormous number of growth periods, bin development depends on the value of F as compared with the value of D. For number-system-like $F = D + 1$ bin schemes, bin proportions are steady and consistent throughout the entire bin structure, lacking development. For fast-expanding $F > D + 1$ bin schemes, bin configurations develop towards skewer configurations, swiftly reaching a particular skewness level, and then maintaining that configuration in all subsequent bin cycles. For slow-expanding $F < D + 1$ bin schemes, bin proportions develop towards less skewed and more equal distributions (an inverse development pattern!) swiftly reaching a milder skewness level, and then maintaining that configuration thereafter in all subsequent bin cycles. Interestingly, here we find that the particular relationship $F = D + 1$ has another significance in a context totally outside the realm of number system construction! For deterministic data, only bin expansion of the exact $F = D + 1$ form yields bin proportions that are steady and consistent throughout the entire bin structure without any bin development.

Chapter 70

Tugs of War between Addition and Multiplication of Random Variables

The sum (addition) of numerous independent realizations from a random variable leads to the symmetrical Normal Distribution. Normal $= X_1 + X_2 + X_3 + \cdots + X_N$, where X_I are independent realizations from an identical random variable X. The Central Limit Theorem (CLT) guarantees Normality for the sum in the limit as N gets large, and (almost) regardless of the distribution form or parameters of X.

The Lognormal Distribution is defined as a random variable whose natural logarithm is the Normal Distribution.

$$\text{Lognormal(location, shape)} = e^{\text{Normal(location, shape)}}$$

$$\text{Normal(mean, s.d.)} = \text{LOGe(Lognormal(mean, s.d.))}$$

This definition together with the result of the CLT imply that the Lognormal distribution can be represented as a process of repeated multiplications of a random variable, namely as:

$$\text{Lognormal} = e^{[\text{Normal}]} = e^{[x_1 + x_2 + x_3 + \cdots + x_N]}$$

$$\text{Lognormal} = e^{x_1} e^{x_2} e^{x_3} \cdots e^{x_N}$$

This result, namely that repeated multiplications of a random variable is Lognormal in the limit as N gets large is called the Multiplicative Central Limit Theorem (MCLT).

As was shown earlier in the book in Chapter 33 when the focus was on quantitative configurations, multiplication processes favor the small over the big, leading to skewed data.

Random multiplication processes induce two essential results:

(A) A dramatic increase in skewness — an essential criterion for Benford behavior.
(B) An increase in the order of magnitude — another essential criterion for Benford behavior.

The resultant Physical Order of Magnitude (POM) of multiplication of several (distinct or identical) random variables is simply the product of the POMs of each of the participating variables. For example, for the random multiplicative process of $X \times X \times X \times Y$, where variable X is of 2 POM, and variable Y is of 5 POM, resultant POM of the entire random product of the four variables is simply $\text{POM}_{\text{PRODUCT}} = 2 \times 2 \times 2 \times 5 = 40$. By definition POM > 1, therefore it always increases with each extra participating multiplicand (variable). In concise mathematical notations, resultant POM of the product of several participating random variables is:

$$\text{POM}_{\textbf{PRODUCT}} = \prod \text{POM}_{\textbf{J}}$$

Surprisingly, multiplication process of many random variables, each with very low order of magnitude, is neither quantitatively skewed nor digitally Benford, but rather quite symmetrical, while the histogram of the process resembles the Normal Distribution a great deal. Here, MCLT is still applied, since there are many variables involved, and it still guarantees that resultant distribution is Lognormal, albeit with low shape parameter. This possibility renders many such types of multiplication processes non-Benford. In stark contrast, the multiplication of only 2 or 3 random variables, each with extremely high order of magnitude — where MCLT does not apply for lack of sufficient number of multiplied variables — still yields nearly the Benford digital configuration and quantitatively skewed distribution, yet resultant data is not Lognormal. These two odd cases teach us that in extreme generality:

(1) MCLT does not guarantee Benford behavior in multiplication processes.
(2) MCLT is not necessary for Benford behavior in multiplication processes.
(3) Order of Magnitude is the crucial factor in determining Benford behavior in multiplication processes.
(4) The confluence of two factors leads to high resultant POM and thus to Benford and GLORQ behavior and skewness in multiplication processes: (a) POMs of the individual multiplied variables, (b) the number of variables that are being multiplied.

Random addition processes do not induce any results that are essential in the criterion for Benford & GLORQ behavior, and it can be said in extreme generality that addition processes favor the medium over the small and over the big. The CLT predicates that the Normal distribution is obtained in repeated additions of any random variable.

Random addition processes are:

(A) Lacking in any increase in skewness, and even actively increasing the symmetry of resultant distribution, with added concentration forming around the center/medium.
(B) Lacking in any increase in order of magnitude beyond the existing maximum order of magnitude within the set of added variables.

The resultant Physical Order of Magnitude of the addition of several *identical* random variables is simply the POM of any one of the individual variables, namely that:

$$\textbf{POM}_{\textbf{SUM}} = \textbf{POM}_{\textbf{THE IDENTICAL VARIABLE}}$$

The resultant Physical Order of Magnitude of the addition of several *distinct* random variables is constrained in being less than or equal to the maximum POM value in the set of all these distinct variables, namely that:

$$\textbf{POM}_{\textbf{SUM}} \leq \textbf{POM}_{\textbf{MAX}}$$

Often in real-life data however, multiplication and addition processes mix together within one measurement or expression, and consequently they fiercely compete for dominance, each attempting to exert the greatest influence upon sizes, quantities, and digits. While addition favors the medium and dislike Benford, multiplication prefers the small and is often Benford itself, and none is willing to compromise. As an example relevant to real-life accounting and financial data, the single bill for one typical shopper in a large supermarket or at a big retail store may read as follows:

$$\mathbf{3} \times (\$2.75 \text{ bread}) + \mathbf{5} \times (\$2.50 \text{ tuna}) + \mathbf{2} \times (\$7.99 \text{ cheese})$$
$$= (\$36.73 \text{ total bill})$$

As another example relevant to real-life scientific and chemical data, the weight of a complex chemical molecule is derived from the linear combination of its constituent atoms.

For example, lactose $(C_{12}H_{22}O_{11})$ has the molar mass of 342.29648 g/mol. This particular molecular weight is derived from the combinations:

$$\mathbf{12} \times (\text{Carbon Mass}) + \mathbf{22} \times (\text{Hydrogen Mass}) + \mathbf{11} \times (\text{Oxygen Mass})$$
$$= (\text{Lactose Mass})$$

$$\mathbf{12} \times (12.0107) + \mathbf{22} \times (1.00794) + \mathbf{11} \times (15.9994)$$
$$= 342.29648 \text{ g/mol}$$

Here one multiplicand is the atomic weight in the Periodic Table, and the other multiplicand is the number of atoms per element within the molecule.

The *CLT's Achilles' heel* — in terms of its rate of convergence to the Normal — is the adverse possibility that added variables are highly skewed and come with very high order of magnitude, constituting a very bad combination for the CLT and a challenge that needs to be overcome. Except for Uniforms, Normals, and other symmetrical distributions which converge to the Normal quite fast after very few additions regardless of the value of order of magnitude of added variables, all other asymmetrical (skewed) distributions show a distinct rate of convergence depending on the value of their order of magnitude. For skewed variables, whenever order of magnitude is of very high value, CLT can manifest itself with difficulties, and very slowly, only after a truly large number of additions of the random variables. On the other hand, when skewed variables are of very low order of magnitude, CLT achieves near Normality quite quickly after only very few additions. Fortunately for revenue and expense accounting data as well as for Molar Mass data in chemistry, the CLT's Achilles' heel saves them from deviation from Benford in spite of the fact that some addition terms are involved, and this is due to the fact that price list and catalogs, as well as the Periodic Table, both are skewed and come with relatively high order of magnitude which induces resistance to the CLT and Normality.

Let us summarize the effects of random arithmetical processes on resultant data (excluding the consideration of orders of magnitude, the number of arithmetical applications, and asymmetry):

Randomly multiplying Uniforms yield Lognormal
Randomly multiplying Normals yield Lognormal
Randomly multiplying Lognormals yield Lognormal
Randomly multiplying Benfords yield Benford
Randomly multiplying non-Benfords yield Benford
Randomly adding Lognormals yield Normal
Randomly adding Uniforms yield Normal
Randomly adding Normals yield Normal
Randomly adding non-Benfords yield non-Benford
Randomly adding Benfords yield non-Benford

Addition Processes: Same POM — More Symmetry — CLT — Normal — Anti Benford

Multiplication Processes: More POM — More Skewness — MCLT — Lognormal — Pro-Benford

As a demonstration of arithmetical tugs of wars, we let U represents the Uniform(5, 33) and perform Monte Carlo computer simulations with 35,000 runs for each arithmetical arrangement. The last value on the right side is the SSD value for each process.

U + U	{ 6.3, 19.2, 31.5, 26.8, 13.9, 2.3, 0.0, 0.0, 0.0}	1360.0
UU + UU	{21.1, 5.7, 9.4, 11.4, 12.4, 11.8, 10.5, 9.4, 8.2}	334.4
UUU + UUU	{42.4, 17.1, 6.7, 4.6, 5.2, 5.6, 6.3, 6.1, 5.9}	222.5
UUUU + UUUU	{30.8, 22.4, 14.7, 9.5, 6.7, 5.1, 4.1, 3.5, 3.3}	38.9
UUUUU + UUUUU	{26.0, 17.7, 13.9, 11.2, 8.9, 7.2, 6.0, 5.1, 4.1}	22.3
UUUUUU + UUUUUU	{30.2, 16.2, 11.5, 9.8, 8.3, 7.1, 6.4, 5.6, 5.0}	4.3

Finding addition sleeping at the wheel, being fixed at 2, multiplication then strives hard to win by gradually increasing the number of multiplicands. Finally multiplication is able to achieve the Benford condition, where it manifests itself six times within the last expression versus only two manifestations of addition. Feeling overconfident now and conceited, multiplication then challenges addition to a new tug of war, willing to face 3 addends, given that it is already in possession of six multiplicands:

$$\text{UUUUUU} + \text{UUUUUU} + \text{UUUUUU}$$

$$\{35.8, 16.4, 9.6, 7.3, 6.5, 6.6, 6.5, 5.8, 5.5\} \quad \text{SSD} = 51.9$$

Upon seeing the severe setback in digital configuration where SSD is now over fifty, multiplication deeply regrets its previous offer, and it then asks addition for permission to use two more multiplicands, achieving a total of eight multiplicands versus these three addends:

$$\text{UUUUUUUU} + \text{UUUUUUUU} + \text{UUUUUUUU}$$

$$\{27.2, 18.5, 13.7, 10.4, 8.5, 6.7, 5.6, 4.8, 4.6\} \quad \text{SSD} = 12.1$$

Multiplication is now quite satisfied with this latest improvement in Benfordness, while addition is not much disturbed upon seeing that digits are getting back fairly close to Benford. Addition knows that ultimately CLT is on its side, and that it would eventually win the war between it and multiplication in due time, and especially when order of magnitude is low or when added distributions are symmetrical and Normality is achieved fairly quickly. Nonetheless, addition wishes to make one last stand in order to flaunt its clout and demonstrate its ability to control the situation, and it expands addends by just one more, namely a total of four addends versus these eight multiplicands:

$$\text{UUUUUUUU} + \text{UUUUUUUU} + \text{UUUUUUUU} + \text{UUUUUUUU}$$

$$\{25.3, 15.5, 13.7, 11.3, 9.5, 7.9, 6.3, 5.5, 4.8\} \quad \text{SSD} = 35.8$$

Indeed, by adding one more addend to the process, addition was able to move it decisively away from Benford, managing to increase SSD from 12.1 to 35.8.

Where would all these actions and reactions, attacks and counterattacks; that endless tit-for-tat war of attrition between addition and multiplication lead to? Without a doubt, the CLT is decisively on the side of addition, guaranteeing its eventual triumph. Surely multiplication could win some battles in the short run, but it would lose the war in the long run as resultant data finally becomes Normally distributed and digital configuration turning decisively non-Benford, and regardless of the form and defined range of added distributions.

Typical Random Linear Combinations (RLC) models in Kossovsky (2014) chapters 16 and 46 are of the particular form:
[Price List] × Dice, or:
[Price List] × Dice1 + [Price List] × Dice2.

RLC derives its Benford tendency exclusively due to the multiplicative nature involved, namely, the product of the Price List by the Dice, and often this is so in spite of the additive terms involved, therefore the Benford behavior of RLC is in harmony and consistent with all that was discussed in this chapter. Indeed, RLC is <u>not</u> any generic or new explanation of the phenomenon of Benford's Law in and of itself in any sense; rather RLC is just another manifestation of the ability of the multiplicative process to serve as a generic and authentic explanation of the Benford phenomenon in many real-life cases.

Surely Price List could be structured as Benford in and of itself, but that's an exogenous issue. And clearly, for a presumed shopper determined to purchase numerous distinct items, say 4 items as in [Price List] × Dice1 + [Price List] × Dice2 + [Price List] × Dice3 + [Price List] × Dice4, revenue data may not be Benford if Price List is with exceedingly low order of magnitude or if it's symmetrical. This is so because the final bill of such 4-item purchase is structured more as additions than as multiplications, and such state of affairs might let CLT ruin any chance toward Benfordness by pointing to the Normal as the emerging distribution. Benford behavior could still be found here for the 4-item purchase, or even for 5 or 6 items, and so forth, given that Price List is highly skewed <u>and</u> that its order of magnitude is very large, because in such cases the Achilles' heel of the Central Limit Theorem would likely prevent the process from achieving any rapid convergence to the Normal distribution. One should keep in mind that even in cases where the Price List is totally symmetrical, and even though Dice is indeed symmetrical, yet their product [Price List] × Dice is skewed and non-symmetrical as in all multiplication processes, and therefore if such skewness is strong enough, and if Price List and Dice combine to yield high order of magnitude for their product, then it might invoke the Achilles' heel of the CLT, prevents the Normal from emerging, and leads to Benford behavior even for a shopper buying 4, 5, or 6 items.

In addition, it should be noted that the claim made on page 191 in Kossovsky (2014) that not only [General Store Shopping] but also [Car] is Benford under certain conditions, supposedly because both processes generate enough variability in resultant data, cannot be valid, unless Car's components are exceedingly skewed and come with some extremely large orders of magnitude, so that the process is resistant to CLT by way of the Achilles' heel of the CLT. Surely, such scenario for [Car] is very unlikely and not realistic at all. In reality, [Car] has to overcome the big hurdle of having so many additive terms within its expression, rendering it highly vulnerable to CLT's tenacious drive towards Normality.

Note: A significant qualification or warning must be added to the four points (1) to (4) on top of page 315, since order of magnitude and POM often suffer from outliers and edges issues, as they are not robust measures of variability of data, thus the most effective measure of the variability of the resultant multiplied data should be Core Physical Order of Magnitude instead (CPOM, namely the 90th percentile divided by the 10th percentile). This is so because here even a large sum of orders of magnitudes of multiplied variables may still be irrelevant as Log(CPOM) of resultant multiplied data could be found with values well below 3.0 or even below 2.0. An illustrative example of this may be given by Uniform(1, 10000) × Uniform (1, 10000), with $4 + 4$ or 8 as the sum of the orders of magnitudes, while Log(CPOM) of resultant multiplied data is actually only around 1.4, hence the resultant multiplied data is indeed not close enough to the Benford configuration.

Chapter 71

Quantitative Partition Models and Benford's Law

Random Real Partition where the pipe or generic quantity is thoroughly broken having plenty of resultants parts is nearly Benford, but partition with only few resultant parts is not. A partition with approximately over 5,000 or 10,000 parts is always nearly Benford, resulting in SSD at around the 8 to 10 level, but it never gets even closer to Benford no matter. Nothing or not much is gained by increasing the number of parts from this level of 5,000 to 10,000, since beyond around this level saturation sets in. In one Monte Carlo computer simulation, a pipe of 800-meter length was partitioned via the Uniform(0, 800) by placing 10,000 random marks along the pipe. First digits distribution came out as {32.2, 16.7, 11.3, 8.6, 7.5, 6.8, 6.0, 5.8, 5.1}, yielding 8.2 SSD value, and such low value indicates that this partition is fairly close to Benford. Here there are 10000/800 or 12.5 marks per unit, and therefore this process corresponds to the Exponential Distributions with 12.5 Lambda parameter value. Indeed, in one computer simulation run for this Exponential Distribution, digits came as {32.3, 16.5, 11.4, 8.7, 7.5, 6.8, 6.3, 5.4, 5.1}.

Random Dependent Partition converges to Benford quite rapidly, and after 10 stages, having merely 2^{10} or 1,024 pieces, SSD is almost always below 10. After 14 stages, having 2^{14} or 16,384 pieces, SSD is almost always below 1 and saturation sets in. In one Monte Carlo computer simulation, a rock weighing 33 kilograms is repeatedly broken in 13 stages into 2^{13} or 8,192 pieces by

321

randomly deciding on the breakup proportions via the Uniform(0, 1). First digits distribution came out as $\{29.9, 17.2, 12.6, 9.7, 8.1, 6.7, 5.9, 5.6, 4.3\}$, yielding the SSD value of 0.6, indicating superb compliance with Benford's Law. Further scrutinizing the set of 8,192 randomly obtained pieces after the 13th stage reveals that the original 33-kilogram rock has been thoroughly divided into totally distinct parts, so that the set of 8,192 pieces does not contain any duplicated quantity. In other words, the resultant set of 8,192 pieces contains 8,192 distinct sizes.

Interestingly, as suggested by the mathematician Steven Miller, even a deterministic fixed p ratio [and its complement $(1 - p)$] can be applied, such as in, say, 20%–80%, or 40%–60%, instead of utilizing random ratios via the Uniform(0, 1) distribution. The fixed ratio result is constrained to cases where $LOG((1 - p)/p)$ cannot be expressed as a rational N/D value, N and D being integers. There are two factors which render Miller's deterministic case less relevant to real-life physical data sets. The first factor is the extremely slow rate of convergence in the fixed deterministic case, which necessitates thousands if not tens of thousands of stages, in contrast to the random case which rapidly converges extremely close to Benford after merely, say, 10 or 13 cycles. It may be that there exist some very long deterministic decomposition processes in nature which involve such huge number of stages, but one would conjecture that these must be quite rare in nature even if they exist at all, and that they are not the typical data sets that the scientist, engineer, or statistician encounters. The second factor is the rarity with which decompositions in nature are conducted with such precise, fixed, and orderly ratio, and perhaps this never occurs at all. Mother Nature is known to behave erratically and chaotically when she feels weak and unable to hold her compounds intact anymore, passively letting them decompose slowly and gradually, and be partitioned in a random way, using totally random ratios. She is even more chaotic when she rages and in anger spectacularly explodes her constructs into bits and pieces rapidly in quick successions; and to expect her to deliberately, calmly, and steadily apply continuously the same fixed p ratio is unrealistic. Expecting to find in nature a decomposition process that (1) comes with an enormous number of stages, and (2) that it steadily keeps the same deterministic fixed p ratio throughout, is being doubly unrealistic.

Those who know Mother Nature well and are familiar with the way she works also assert that she would probably never bother to delicately break her quantities carefully according to fixed stages, and this line of thought suggests that also Random Dependent Partition is not the typical process that she likes doing or even capable of performing, since this process demands some mental effort and organization.

Let us imagine a real-life assembly-line with workers and management as in typical large corporations, attempting to physically perform the Random Dependent Partition process on a very long metal pipe. First the workers decide on the first random location where the pipe is to be cut, followed by actual cutting, and then the designations of "1" and "2" tags for the newly created pieces are made. This is followed by the orderly cutting of piece "1" and then piece "2" using random location within each piece. Without such designations and tags, there exists the possibility that by mistake one piece is cut twice leaving the other piece intact. Next, the workers designate the newly created four pieces as "1", "2", "3", and "4". At this stage, piece designation becomes even more crucial to avoid confusion and mistakes, and to remember which pieces were already cut and which pieces are awaiting their turn. It is highly doubtful that Mother Nature is capable or even interested in such serious and rigid type of work.

How would temperamental Mother Nature go about breaking a rock or a pipe her way, leisurely, chaotically, and consistent with her free-spirit attitude and her strong dislike of regimentation?

Her first act in the process is the breaking of the original rock into two parts randomly via the continuous Uniform(0, 1) to decide on the proportions of the two fragments. There is no need whatsoever to designate any pieces with any tags thereafter, since order of breakups does not matter to her in the least. Her second act is the totally relaxed and random selection of any one of the two pieces, followed by its fragmentation into two parts randomly via the continuous Uniform(0, 1), resulting in three pieces. Her third act is the totally relaxed and random selection of any one of the three pieces, followed by its fragmentation into two parts randomly via the continuous Uniform(0, 1), resulting in four pieces. If this continues for sufficiently large number of stages [approximately 10,000 or 20,000] then the

Benford configuration is obtained nearly perfectly. This process is coined as **Chaotic Independent Partition.**

Three distinct Monte Carlo computer simulations of Chaotic Independent Partition were run, all starting with an initial 100-kilogram rock, having 500, 1300, and 20000 stages, and applying the random breakup ratio of the Uniform(0, 1). The three schemes yielded the following results:

Chaotic Independent Partition 500 Stages:
{32.1, 19.4, 11.6, 8.6, 8.0, 5.8, 6.0, 4.0, 4.6}, SSD = 11.4
Chaotic Independent Partition 1300 Stages:
{29.0, 18.6, 12.8, 9.4, 9.8, 6.1, 5.6, 4.7, 4.1}, SSD = 6.6
Chaotic Independent Partition 20000 Stages:
{30.0, 17.2, 12.4, 9.9, 8.0, 6.8, 6.0, 5.1, 4.8}, SSD = 0.3
Benford's Law for the First Digits:
{30.1, 17.6, 12.5, 9.7, 7.9, 6.7, 5.8, 5.1, 4.6}, SSD = 0.0

In extreme generality, the Benford digital configuration is found in quantitative partition models whenever these partitions are performed in a purely *random fashion* and are done repeatedly enough so that the original quantity is finally broken into *numerous resultant parts*. The adherence to the first of these two requisite features, namely the partitioning in a random fashion, almost always guarantees two other consequences that are also essential for achieving the Benford configuration, namely breaking the original quantity *along the real number basis* and not exclusively along the integers, as well as ensuring that the original quantity is broken into (mostly) *distinct resultant parts* with almost no repetitions of sizes.

A **Consolidations and Fragmentations Process** (C&F) is a process where an initial large set of identical quantities (say balls) constantly alternates between random consolidations (the fusing of two randomly chosen balls) and random fragmentations (the splitting up of one randomly chosen ball via randomly chosen ratio). This process converges eventually to the Benford configuration after sufficiently many such cycles. This cyclical process alternates constantly between the selection of two randomly chosen balls and their summation [consolidation or merge] into a singular fused and bigger ball, and the throwing of this newly created and consolidated ball back into the main pile of balls; followed by the selection of one ball chosen at random and its breakdown [fragmentation] into two smaller balls via the random split as in the format {Uniform$(0, 1)$, 1 − Uniform$(0, 1)$}, and the throwing of these two newly created balls back into the main pile of balls.

This consolidations and fragmentations model is based on L balls, all with an identical initial real value V, so that the initial set is {$V, V, V, \ldots L$ times $\ldots V, V, V$}. The focus here is on the weights of the balls, and assuming uniformity of mass density, implying equivalency between volume and weight proportions. The specific description of physical balls of uniform mass density being fused and then broken, and the focus on the weight variable, is an arbitrary one of course, and the generic model is of pure quantities and abstract numbers.

This C&F model consists of C cycles. Within one full cycle, two opposing processes are performed, one of consolidation, followed by one of fragmentation. Obviously, after each full cycle, the number of existing balls is unchanged, and it is still L, being the same as in the beginning and as in the end of the entire process. Also, the total quantity of the entire system — namely the overall weight of all the balls — is conserved throughout the entire process.

Monte Carlo computer simulations results decisively show that C&F process leads to the Benford digital convergence and to the GLORQ-like quantitatively skewed set of balls. Exploration of the resultant (randomly generated) algebraic expressions coming out of the corresponding mathematical model, nicely explains this Benford behavior simply in terms of a random multiplication process, and this is so in spite of the addition terms involved here. In other words, the arithmetical model of the process points to a tug of war between addition and multiplication, and where multiplication decisively triumphs over addition.

Monte Carlo computer simulations show that after C full cycles, the weights of the balls are very nearly Benford and are highly skewed quantitatively, given that $C > 2 \times L$ approximately, although $C > 3 \times L$ or $C > 4 \times L$ usually give slightly better results. It is necessary to have a sufficient number of these C cycles so that all, or almost all of the balls experience either fragmentation or consolidation (preferably both, and hopefully not merely once, but rather twice or three times). By cycling at least twice as many balls that exist in the system, we ensure that (almost) all the balls undergo transformation of some sort, and that the initial value V is (almost) nowhere to be found among the balls at the end of the entire process. Continuing beyond the required $2 \times L$ or $3 \times L$ cycles [i.e. saturation level] does not ruin the decisive Benford convergence thus obtained, and Benford is steadily preserved (or rather further perfected a bit) as more cycles are added. Surely, the other essential prerequisite for Benford convergence here is to have a sufficiently large number of balls in the system so that the Benford configuration can be properly manifested. Hence the requirement is that $L > 200$, or for better convergence that $L > 300$. Falling below 300 or 200 balls for example yields only crude or approximate Benford-like results, as in all small data sets aspiring to obey the law. Quantitatively though, even for a scheme with small L value, resultant set of balls is almost always decisively skewed in favor of the small.

For the Consolidations and Fragmentation model, each fragmentation process contributes to the system two products, namely Ball $\times$ Uniform$(0, 1)$ and Ball $\times$ $(1 -$ Uniform$(0, 1))$, each with a minimum of two multiplicands and possibly more; similarly each consolidation process contributes to the system one additive expression, namely $\text{Ball}_1 + \text{Ball}_2$, with a minimum of two addends and possibly more; and therefore there exists a tug of war here between additions and multiplications with respect to Benford behavior. One existing feature here that partially saves the system from deviation from Benford is that on average only about one-third of the expressions are additive, while about two-third of them are multiplicative; and that even within those additive expressions there are plenty of arithmetical multiplicative elements involved. Surely there are some additive expressions with three addends which might be quite detrimental to Benford, but they are far and few between; and there are even more menacing expressions with four addends, but luckily these are even rarer.

The general theoretical understanding gained in the previous chapter regarding tugs of war between multiplication and addition enables us to thoroughly explain the empirically found decisive Benford behavior in C&F models; namely the reason addition effects do not manage to significantly retard multiplication effects. In a nutshell, the C&F process is Benford because it uses the high order of magnitude variable of Uniform$(0, 1)$ which contributes to the process itself as a whole high order of magnitude for the resultant set of balls;

while skewness is guaranteed via the involvement of multiplication. As a consequence, the C&F process encounters the Achilles' heel of the Central Limit Theorem and so additions are not very effective. Order of magnitude of the Uniform(0, 1) calculated as LOG(1/0) is infinite. CPOM calculated as $P_{90\%}$ divided by $P_{10\%}$ is $0.9/0.1$ or simply 9. Surely, the C&F model cannot use any low order of magnitude variable such as say Uniform(5, 7), because it needs to break a whole quantity into two fractions, and this can only be achieved via the high order of magnitude variable of Uniform(0, 1). Such high order of magnitude values, coupled with the fact that the terms within the additive expressions almost always involve also some multiplications (which are always skewed), guarantee that the Central Limit Theorem is very slow to act here and that its retarded rate of convergence does not ruin the general multiplicative tendencies of the system in the least.

Surprisingly, deterministic models of fixed 50%–50% even ratios, and especially fixed uneven ratios such as 15%–85% for example, also converge to Benford. Even though these deterministic models do not apply the Uniform(0, 1) in any way, the resultant expressions of such deterministic p and $(1-p)$ ratios are also of high order of magnitude values, which guarantees sufficiently high order of magnitude for the whole system and prevents the CLT from manifesting itself due to the Achilles' heel of the CLT. Randomness in such deterministic fixed ratio models still exists in the system in the sense that the balls to be consolidated and the balls to be fragmented are chosen in a random fashion, and [in extreme generality] randomness is almost always a necessary condition for Benford and skewness in quantitative partition models.

Chapter 72

Chains of Statistical Distributions Are Nearly Always Benford

Nearly all chains of distributions are Benford or approximately so. In some particular cases this is achieved by having sufficiently large number of distributions involved, namely that the chain is long enough having plenty of sequences, but almost always having merely three or four sequences is sufficient for strong convergence to Benford.

The general statement of Chapter 63 titled "Benford and GLORQ as a Subset of the Small Is Beautiful Phenomenon" also applies here for chains of distributions. Hence, even when a particular chain of distribution does not converge to Benford, the small is beautiful phenomenon is almost always still valid, albeit with different quantitative configurations as compared with that of the Benford type, and typically of a milder form of skewness.

Order of magnitude of chains of distributions normally does not factor directly or obviously in convergence to Benford, although it does play a hidden role indirectly in facilitating convergence. Hence, when digital results from a wide variety of chains of distributions (of similar format and style but with distinct ultimate parameters) are compared, it often shows a strong correlation between the chain's order of magnitude and its convergence to Benford.

As an example, the four chained Uniform Distributions of Chapter 41 Figure 2.32, with 10,000 simulation runs, shall be digitally analyzed. Formally written, this chain scheme is expressed as **Uniform(0, Uniform(0, Uniform(0, Uniform(0, 55))))**. The start at the 0 origin allows for the use of any b parameter value whatsoever, not merely for the "big" value of 55, but also for very "small" values such as 0.0008 say, with the same resultant success in convergence, and without worrying about the order of magnitude at all. In principle, the start at 0 guarantees an infinite order of magnitude for the chain, although when outliers and edges are taken into consideration — as in the definition of CPOM — the start at 0 only guarantees sufficiently high order of magnitude (albeit finite) for excellent convergence. First digits distribution for this chain of four Uniforms is {30.6, 17.5, 11.8, 9.3, 8.4, 7.0, 5.9, 5.1, 4.5}, and the extremely low 1.3 SSD value indicates that this chain is nearly perfectly Benford. The minimum value here came out as 0.0000154, and the maximum value came out as 41.4, therefore POM = (41.4)/(0.0000154) = 2689814, while OOM = LOG(2689814) = 6.4. This might seem to be quite high, yet outliers and edges here are at fault for exaggerating the value of the order of magnitude. The 10% percentile here is 0.066, and the 90% percentile is 9.552, hence CPOM = (9.552)/(0.066) = 144.5.

The chain **Exponential(Exponential (Exponential(7)))** is another example with three Exponential Distributions, and which converges to Benford nearly perfectly. In 10,000 simulation runs, the first digits distribution came out as {30.3, 18.1, 12.3, 10.2, 8.0, 6.5, 5.1, 5.0, 4.4}, and the extremely low 1.2 SSD value here indicates that this chain is nearly perfectly Benford.

The chain **Exponential(Normal (Uniform(15, 21), 3))** is an example of a hybrid chain composed of three distinct distributions, and which converges to Benford as expected. In 10,000 simulation runs, first digits came out as {29.6, 16.0, 11.9, 10.0, 8.5, 7.4, 6.1, 5.6, 4.8}, and the very low 4.5 SSD value here indicates that this chain is very close to Benford.

Although the order of magnitude of chains of distributions normally does not factor in an obvious way regarding convergence to Benford, one exception (among others) to this general statement is the short chain of the form **Uniform(A, Uniform(A, B))**, $B > A$, where convergence to Benford cannot be achieved at all without sufficient order of magnitude of the set $\{A, B\}$, namely having sufficiently large ratio of B/A of at least 10 approximately. Mysteriously, the exact ratio of 13 for B/A for some reason yields the best result here, having the strongest compliance with Benford. Surpassing the value of 13 yields worsening results with wider deviation from Benford, even though order of magnitude increases when the ratio of B/A is made larger.

For example, **Uniform(3, Uniform(3, 6))** is of extremely low order of magnitude as seen in the ratio $B/A = 6/3 = 2$, and clearly the only values this chain generates are between 3 and 6 exclusively, hence digits $\{3, 4, 5\}$ are the only possible first digits, while digits $\{1, 2, 6, 7, 8, 9\}$ never even occur on the first place.

The opposite pole of the above unsuccessful example is given by the superior chain **Uniform(3, Uniform(3, 9000))** which is of extremely high order of magnitude as seen in the ratio $B/A = 9000/3 = 3000$. Clearly, the values this chain generates are widely spread between 3 and 9000. In one specific Monte Carlo computer simulations with 20,000 runs, digit distribution came out as $\{25.6, 18.8, 14.6, 11.9, 9.3, 7.2, 5.5, 4.1, 3.0\}$, with a moderate SSD value of 36.9. The lowest value obtained was 3.03, and the highest value obtained was 8929.89, resulting in POM value of $8929.89/3.03 = 2944.3$. The 10th percentile obtained was 193.9, and the 90th percentile obtained was 5310.8, resulting in CPOM value of $5310.8/193.9 = 27.4$.

Another remarkable convergence to Benford for this very short chain of only two sequences can be found in the following case where the ratio $B/A = 13$, and which yields an extremely good fit: **Uniform(17, Uniform(17, 221))**, $\{30.2, 17.1, 11.8, 9.7, 8.0, 7.0, 6.0, 5.4, 4.9\}$, and SSD = 1.1.

The chain **Uniform(0, Uniform(0, B))** on the other hand theoretically possesses an infinite order of magnitude for any value of B. This is so since $B/0 = \infty$, and in fact its partial convergence to Benford is accomplished with any B value whatsoever. Nonetheless, the degree of convergence to Benford depends on the value of B, with SSD values [cyclically] fluctuating between 25 and 55. Surely when the thorny issue of outliers and edges is taken into consideration, and the CPOM measure is applied, the true variability of the chain is understood to be finite, and not even very large, all of which explains why the chain does not fully converge to Benford, but only partially so.

The author's first conjecture is that an infinitely long chain of distributions should obey Benford's Law exactly. For more discussions and other related conjectures regarding chains of distributions see Kossovsky (2014) Chapters 54, 102, and 103.

The following is a very brief summary in extreme generality of these three chapters:

Scale parameters such as λX or X/λ (divisions and multiplications) as well as Location parameters such as $X - \mu$ (subtractions), usually respond vigorously to chaining, prefer the small over the big, and obey Benford's Law (i.e. chain-able). Shape parameters such as X^k (powers) usually do not respond to chaining at all, show no preference for the big, for the small, or for any size, and disobey Benford's Law (i.e. not chain-able).

More precisely, a parameter that does not continuously involve itself in the expression of centrality [such as the mean, median, or midpoint] is not chain-able at all; and a parameter that does continuously involve itself in the expression of centrality is indeed chain-able.

The meaning of the phrase "not continuously involved" implies that the partial derivative $\partial(\text{center})/\partial(\text{parameter})$ goes to 0 in the limit for high values of the parameter, namely that centrality such as the average, the median, and resultant range in general, are not affected as parameter is further increased; that beyond a certain limit the parameter does not sway centrality.

$$\lim_{\text{parameter}\to\infty} \partial(\textbf{center})/\partial(\textbf{parameter}) = 0$$

Hence, if a parameter does not play any role in the determination of centrality and span of range beyond the initial few low values then it's not chain-able at all. Since most scale and location parameters [continuously] play significant role in centrality, they are generally chain-able. Since most shape parameters do not play any role in centrality, they are generally not chain-able.

In addition, a second conjecture in made, predicting the manifestation of Benford's Law exactly for the very short chain of distributions with even just two sequences, assuming the chain uses a distribution which obeys Benford's Law for the inner-most parameter in its ultimate sequence. As an example of the second conjecture, the chain **Uniform(0, Exponential Growth Series)** is predicted to closely obey Benford's Law, even though it only has two sequences. This is so since Exponential Growth Series is in and of itself quantitative skewed and it obeys Benford's Law almost perfectly.

In extreme generality and concisely in symbols, the second conjecture states that:

Any Distribution(Any Benford) = Benford

A related extrapolation of the second conjecture states that with each new added sequence (elongating the chain), the chain evolves and becomes even skewer, as well as becoming a notch closer to the digital configuration of Benford's Law.

Chapter 73

Development Pattern within the Expanding Histograms of Benford

This chapter illustrates visually the Digital Development Pattern as embedded within the structure of the bin scheme of GLORQ in the context of a real-life numerical example applied to Benford's Law.

Accordingly, we shall demonstrate visually the confluence of two essential features in Benford's Law, namely the GLORQ interpretation of Benford's Law as an expanding set of histograms, together with the Digital Development Pattern. Facilitating this demonstration is the fact that first digits distribution of a given data set is nothing but the condensed or aggregated histogram of the various nine-bin local histograms standing between integral powers of ten such as 1, 10, 100, 1000, and so forth.

The data on US population regarding all its 19,509 cities and towns in the 2009 census survey is the real-life numerical example chosen to demonstrate the interactive effects of these two essential features in the field. Ignoring the largest nine mega cities with over one million inhabitants (considered as outliers), we are left with 19,500 population centers. Figures 4.9–4.14 depict the six local histograms constructed on the relevant ranges standing between IPOT values, from the minimum of 1 all the way to the maximum of 1,000,000.

The Digital Development Pattern is now visually and clearly demonstrated for this data set. On the first two sub-intervals (1, 10) and (10, 100) the histograms are actually on the rise, and the big is more numerous than the small locally. Yet, since very little data

1 - 10

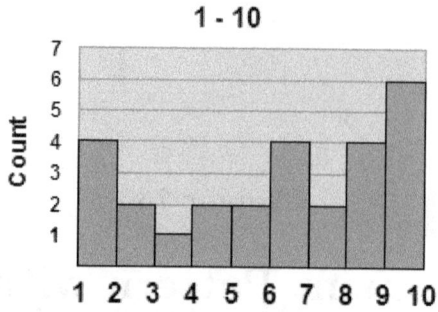

Figure 4.9:   US Population on (1, 10)

10 - 100

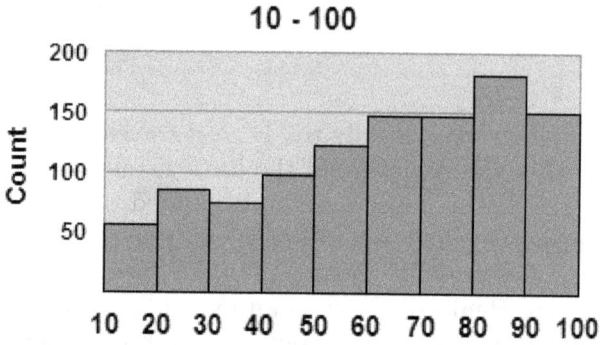

Figure 4.10:   US Population on (10, 100)

100 - 1,000

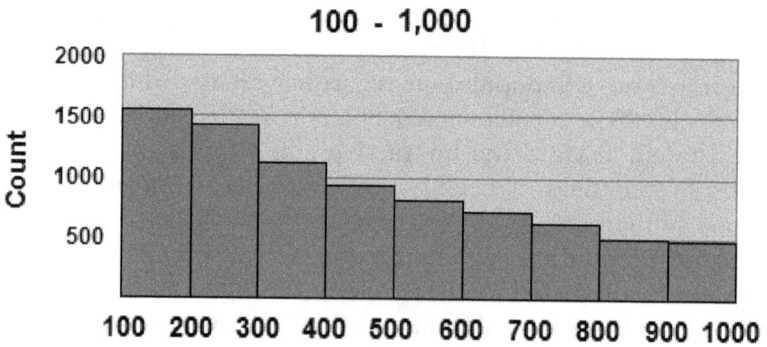

Figure 4.11:   US Population on (100, 1,000)

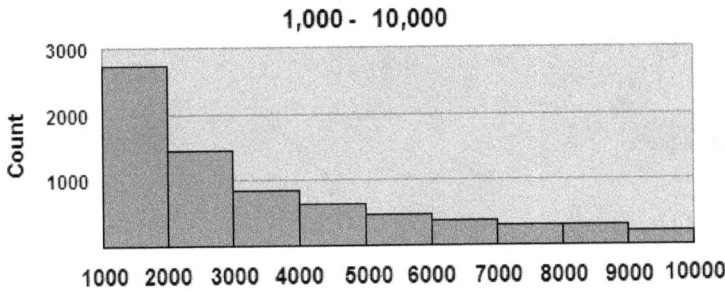

Figure 4.12: US Population on (1,000, 10,000)

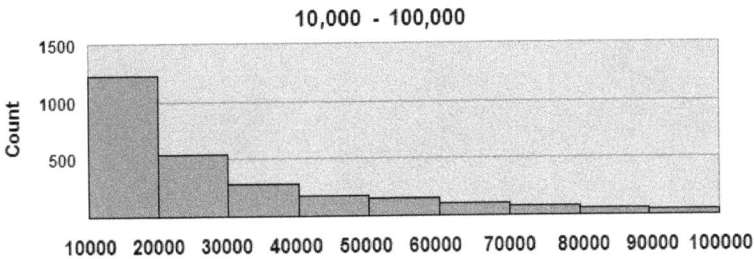

Figure 4.13: US Population on (10,000, 100,000)

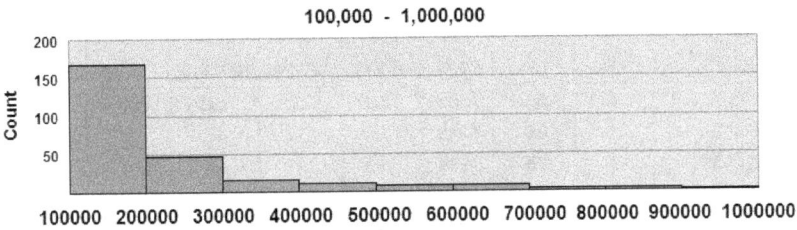

Figure 4.14: US Population on (100,000, 1,000,000)

portion falls within these two sub-intervals, the small is beautiful phenomenon and Benford's Law can easily manifest themselves in the aggregated data over all sub-intervals. Further to the right on (100, 1,000) the histogram is falling gently and the small is only mildly more numerous than the big within this locality. Finally, much further to the right from 1,000 to 1,000,000 the histograms are precipitously falling, severe skewness is observed, and the small is extremely more numerous than the big, much more so than in the Benford configuration.

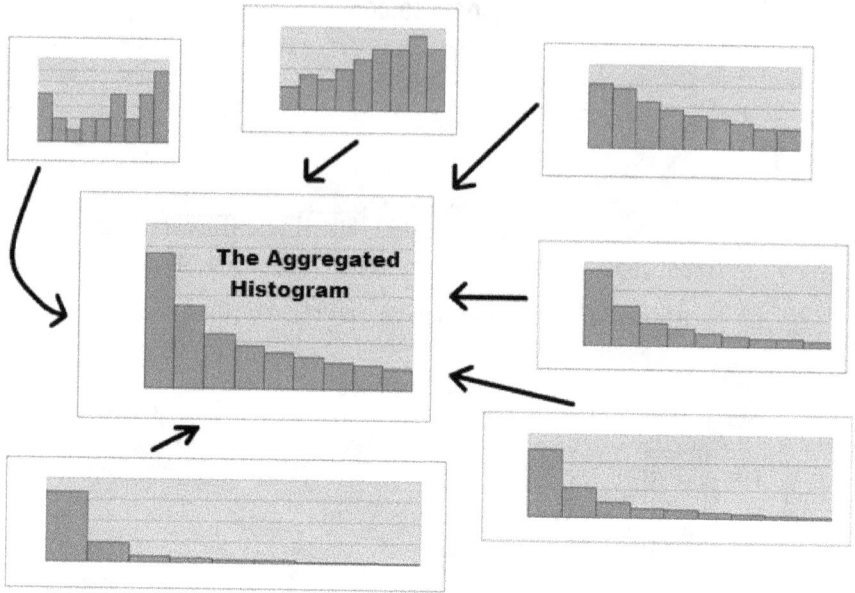

Figure 4.15:   The Fusion of Local Histograms into an Aggregated Histogram

From :	1	10	100	1000	10000	100000	All Data	All Data
Up to:	10	100	1000	10000	100000	1000000	Count	Proportion
Bin 1	4	56	1565	2718	1222	168	5733	29.4%
Bin 2	2	86	1429	1437	537	47	3538	18.1%
Bin 3	1	75	1116	843	290	16	2341	12.0%
Bin 4	2	98	941	624	171	11	1847	9.5%
Bin 5	2	123	813	460	153	8	1559	8.0%
Bin 6	4	148	721	388	101	8	1370	7.0%
Bin 7	2	148	626	311	75	4	1166	6.0%
Bin 8	4	181	502	292	60	3	1042	5.3%
Bin 9	6	150	489	212	45	2	904	4.6%
Number of Cities	27	1065	8202	7285	2654	267	19500	19500
Data Proportion	0.1%	5%	42%	37%	14%	1.4%	100%	100%

Figure 4.16:   Table of US 2009 Population Data by Local Sub-Intervals

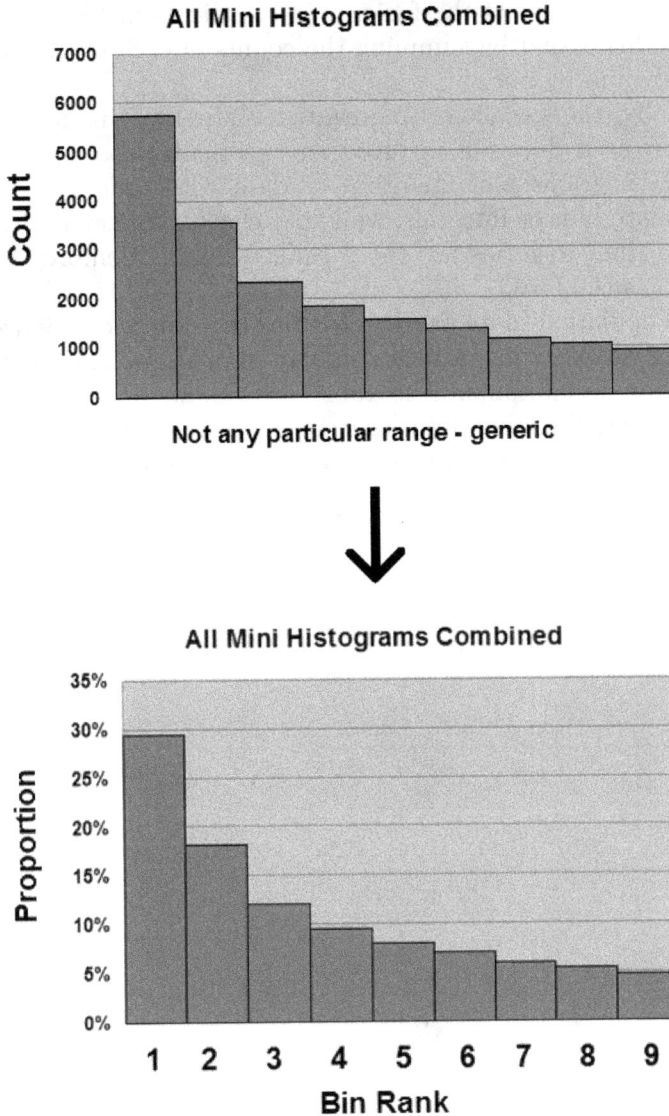

Figure 4.17:   The Final Step: Conversion of Counts into Proportions

Note: Sizes are not shown properly in their true exact proportions — for lack of space. There are actually 10-fold expansions in the width between successive local histograms, and such dramatic expansions are just too difficult and space-consuming to show here.

Figure 4.15 depicts the fusion of these six local histograms into one large histogram by summing the counts of each of the nine bins separately, bin by bin.

Actually, the so-called "Aggregated Histogram" is indeed just a bar chart, as it does not involved any particular sub-ranges on the x-axis. Histograms plot quantitative data with ranges of the data grouped into bins or intervals, while bar charts plot categorical data, and here the categories are the 9 ranked digits (Benford) or the 9 ranked bins (GLORQ).

US Population data is detailed in the table of Figure 4.16, excluding those nine outliers, namely the mega cities with over one million inhabitants. The table also shows the conversion of the aggregated counts into aggregated percentages. Figure 4.17 depicts conceptually the final step of the conversion from counts into percentages.

Appendix A

Conceptual Explanation
of Histograms

Often we come across a group of many related numbers representing the same phenomenon, such as for example the weights in kilogram of all the students at a large university in prosperous Sweden of Northern Europe, where the food is rich and varied, and the genetic makeup of the natives is of solid, tall, and strong body frame. We may wish to compare this Swedish data set to another set of weights in kilogram of all the students at another university located in Venezuela of South America, with fewer resources and less adequate nutrients due to political difficulties, and with an overall genetic makeup tendency for relatively thin and modest body frame. Surely, the overall descriptions of these two data sets would be starkly different from each other. It is necessary to think and plan ahead as to how best one should present and summarize each distinct data set and capture its inner numerical and quantitative structure. Clearly, simple-minded calculations of the two different averages or of the two distinct ranges from the minimum to the maximum are not sufficient.

In another example, we may wish to study the blood cholesterol level of the entire population of a remote tribal village of hunters in the Amazon jungle basin somewhere in northern Brazil, eating natural and freshly hunted unprocessed meat, together with plenty of green vegetables, colorful fruits, and bitter roots, while staying physically active daily. Biologists would naturally attempt to compare this cholesterol data set to another set of say the entire population of New York City, where the food is highly processed and not as healthy, often consisting of high calorie intakes such as cheeseburgers, chocolate, ice cream, and cakes, and where people often lead more sedentary lives, relying mostly on subway trains, buses, and taxis, to move around, especially in the harsh and bitterly cold winters.

With our keen eyesight, we can easily examine and analyze these numbers on a computer screen or printed paper. Our good vision and intellectual capacities to process and analyze information enable us to summarize and compare distinct data sets. We should appreciate our ability to see and experience the numbers visually, unlike those who are blind and depend on listening to detailed verbal descriptions of numbers for analysis, followed by their superhuman effort to synthesize and analyze the set of spoken numbers.

Yet, most normal mathematicians, statisticians, and data analysts with good vision, are not satisfied with merely staring at the data with their healthy eyes, they do not count their blessings, and they thrive to arrive at an even better and superior vision that would elegantly summarize the entire data via a singular picture, or rather via what they term "histogram", and which would hopefully stand for all these many numbers in one fell swoop. They often refer to the motto or slogan "A picture is worth a thousand words", which they extrapolate to their self-styled new slogan "A histogram is worth a thousand numbers". This famous phrase is more eloquently and more correctly expressed in Spanish as "Una imagen vale más que mil palabras", converting it from the limited exact equality and into generic inequality, and which could then be extrapolated and expressed in our context as "Un histograma or un gráfico vale más que mil números". Actually, these professionals with full vision should not be blamed or criticized for trying to summarize sets of numbers as such, because at times they need to stare at hundreds or even thousands of pages full of numbers, tiring their eyes, while they are tasked with making sense out of it all, and are asked to report on all of this concisely, so naturally a singular histogram is the best solution for them. The aim is to capture the entirety of all these many numbers and condense them into a singular visual image which would aid us in understanding the quantitative structure and the relative occurrences of all the distinct values within any given data set.

Figure A.1 depicts a small-size data set of cholesterol values taken from 160 randomly chosen inhabitants of New York City. Current medical dogma holds that total cholesterol in adults should be less than 200; borderline to moderately elevated levels are considered to be in the range of 200–240; and any reading over 240 is considered to be problematic and in need of some remedial action, such as better diet, increased physical activity, or medical prescription of cholesterol-lowering drugs. By merely scanning quickly with our eyes this numerical table of Figure A.1, even in a very superficial manner, it is clearly observed that unfortunately most New Yorkers are with either slightly or significantly cholesterol reading of over the 200 threshold level, and that all this does not bode well for the future cardiovascular health of the inhabitants of this city.

The young, inexperienced, and underpaid but enthusiastic novice statistician who has been assigned to scrutinize this medical data set, has just recently earned his master's degree from a prestigious university, and he is now staring long and hard at the numerical table of Figure A.1, attempting to make sense out of all these disparate values, so naturally the task initially overwhelms him. But as soon as he hits upon the idea of sorting the numbers from low to high, he gains confidence, and his ability to analyze the data is now nearly assured. Figure A.2 depicts his newly constructed table of all the numbers sorted from low to high. It is now obvious to him that the range is roughly from 150 to 300, and that therefore the midpoint between these two extremes can be calculated as $(150 + 300)/2$ or simply 225. Moreover, the imminent cardiovascular calamity coming soon to New York City is now very obvious to him, as he observes that only the first column on the left side is of healthy individuals with ideal levels below the threshold of 200, while all the three columns on the right side are with high cholesterol levels of over 200.

Yet, the statistician wants to gain more understanding, and he is curious to know where most of the values occur. Do they concentrate around the center? How many healthy individuals are near the low 150 minimum value? Are there too many unhealthy individuals very near the dangerous maximum value of 300? In other words, what he needs to explore here is the detailed spread of these cholesterol values, and how they are distributed exactly along the long range between 150 and 300. Hence, in order to investigate this, he carefully counts 86 values falling below the midpoint of 225; and he also carefully counts

182.6	201.1	217.1	196.9
151.2	242.9	275.7	182.4
250.7	230.3	249.6	219.8
231.2	223.6	257.1	249.9
150.3	270.5	207.4	230.6
205.8	219.8	207.4	252.1
245.7	238.7	194.2	237.3
259.2	229.5	257.7	256.6
200.7	198.2	245.7	215.4
186.4	161.4	289.5	194.6
164.4	204.6	287.3	202.4
229.3	235.7	174.5	251.7
175.2	242.3	197.7	247.1
255.8	232.7	171.4	213.7
238.4	228.9	184.1	163.7
219.8	228.6	188.5	228.5
243.4	202.5	230.6	208.8
221.6	235.9	173.9	251.8
196.8	215.7	199.1	236.3
222.5	227.2	241.6	275.4
251.9	228.5	176.8	223.6
171.8	214.3	216.4	189.1
210.7	201.9	215.1	171.7
250.6	201.9	224.7	234.7
197.3	202.2	226.8	213.8
179.9	205.1	223.9	225.2
222.3	249.2	203.5	257.9
170.9	218.2	213.6	201.6
230.5	215.3	170.5	263.4
175.3	203.7	236.9	202.8
206.2	240.3	225.8	261.6
207.4	228.4	216.6	225.1
266.1	275.3	199.4	179.1
252.9	225.6	256.7	256.5
239.7	189.1	221.5	223.7
257.2	154.8	233.6	203.3
227.4	201.8	245.3	216.4
189.6	208.6	297.4	231.9
187.6	257.5	204.8	272.5
182.2	267.2	249.4	225.1

Figure A.1:   Cholesterol Levels in New York City — Raw Data

74 values over the midpoint of 225. This reveals to him that the data perhaps is overall symmetrical in nature. Then, in order to emphasize this fact, he decides to present this division between the range of 150 to 225 and the range of 225 to 300 in a visual way by a particular drawing which he calls "histogram", as shown in Figure A.3, where

150.3	201.8	222.5	242.3
151.2	201.9	223.6	242.9
154.8	201.9	223.6	243.4
161.4	202.2	223.7	245.3
163.7	202.4	223.9	245.7
164.4	202.5	224.7	245.7
170.5	202.8	225.1	247.1
170.9	203.3	225.1	249.2
171.4	203.5	225.2	249.4
171.7	203.7	225.6	249.6
171.8	204.6	225.8	249.9
173.9	204.8	226.8	250.6
174.5	205.1	227.2	250.7
175.2	205.8	227.4	251.7
175.3	206.2	228.4	251.8
176.8	207.4	228.5	251.9
179.1	207.4	228.5	252.1
179.9	207.4	228.6	252.9
182.2	208.6	228.9	255.8
182.4	208.8	229.3	256.5
182.6	210.7	229.5	256.6
184.1	213.6	230.3	256.7
186.4	213.7	230.5	257.1
187.6	213.8	230.6	257.2
188.5	214.3	230.6	257.5
189.1	215.1	231.2	257.7
189.1	215.3	231.9	257.9
189.6	215.4	232.7	259.2
194.2	215.7	233.6	261.6
194.6	216.4	234.7	263.4
196.8	216.4	235.7	266.1
196.9	216.6	235.9	267.2
197.3	217.1	236.3	270.5
197.7	218.2	236.9	272.5
198.2	219.8	237.3	275.3
199.1	219.8	238.4	275.4
199.4	219.8	238.7	275.7
200.7	221.5	239.7	287.3
201.1	221.6	240.3	289.5
201.6	222.3	241.6	297.4

Figure A.2: Cholesterol Levels in New York City — Sorted Data

the vertical heights of the two rectangular bins represent the count of each range (i.e. frequency of occurrences).

In his next step he wanted to examine the data further by subdividing the entire range into 3 sections. He then counted 37 low

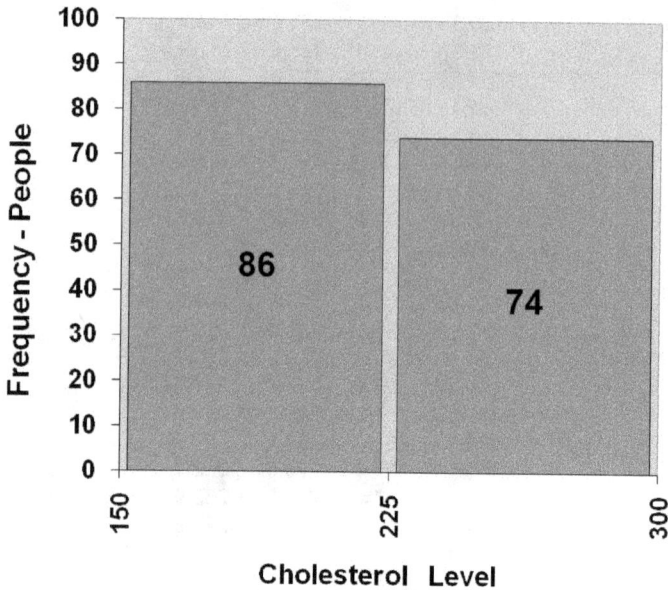

Figure A.3:   Histogram of Cholesterol Levels in New York City — Two Bins

values falling between 150 and 200; 94 medium values falling between 200 and 250; and 29 high values falling between 250 and 300. This more refined scrutiny of the data led to a slightly better histogram as shown in Figure A.4. This new histogram suddenly revealed to him a definite concentration of values around the center. Indeed, the central horizontal line segment between 200 and 250 is by far more "dense" (i.e. having more data points falling in) than either the line segment on the left side of 150–200, or the line segment on the right side of 250–300. This is the origin of the term "density curves", albeit here so far we are dealing only with linear bars and bins and rectangular shapes, as opposed to round curves. This concentration around the center would have been exceedingly difficult to decipher by just staring at the numerical tables of Figures A.1 and A.2. Only the construction of the histogram enables us to notice the concentration around the center as well as the approximate symmetry between the minimum and the maximum. The concept of the "histogram" is based on the concept of "distribution", namely the detailed description of the entire set of numbers in the data and how they are distributed and spread along the horizontal number line from the minimum to

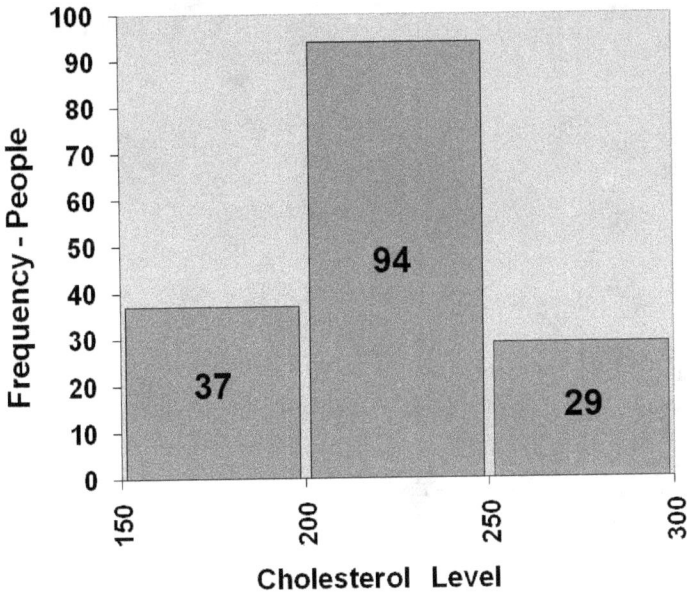

Figure A.4:   Histogram of Cholesterol Levels in New York City — Three Bins

the maximum. In our case, it appears that these 160 cholesterol values are mostly distributed around the center, and are less dense but symmetrically and evenly distributed on the left and right parts of the range. It should be noted that formally there should be no spaces at all between the bins of the histogram, and that the bins should be touching each other. This is so because the construction of the bins involves a partitioning of the entire horizontal axis from the minimum to the maximum points without any breaks or gaps, covering all corners and values. It is only for better visualization and for pedagogical reasons that some non-existent invented spaces appear between the bins of Figures A.3 and A.4, separating them slightly.

What was not yet known about this data set is how values behave within each of the three sections themselves. Are they rising within the central bin perhaps? Are they falling within the right bin? Are they perhaps zigzagging or just approximately flat within the left bin? The enthusiastic novice statistician was determined to answer these questions and to continue to explore proportions of values falling within smaller and smaller segments of the horizontal line.

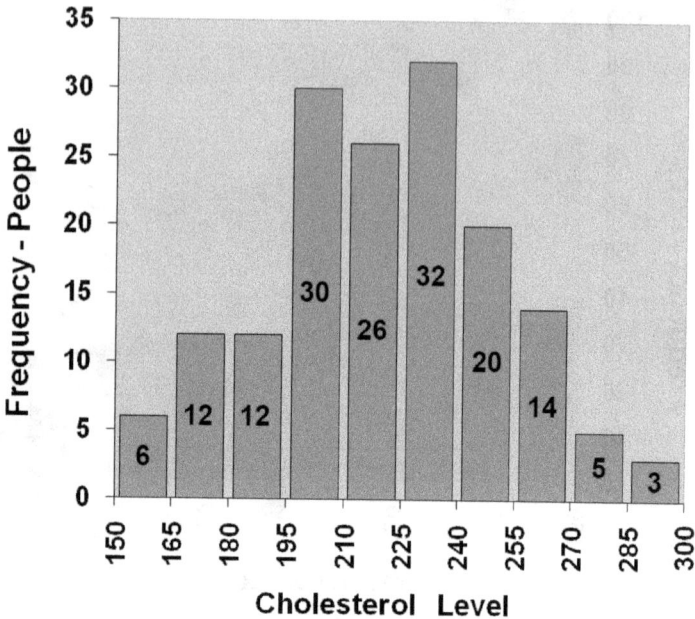

Figure A.5: Histogram of Cholesterol Levels in New York City — 10 Bins

He decided to subdivide the entire range of 150–300 into 10 equal sections, with more refined and shorter steps of 15 cholesterol units each, leading to the constructing of a 10-bin histogram as shown in Figure A.5. Now, with this 10-bin histogram, the statistician has arrived at the best and clearest visual description of this particular cholesterol data set. He now thought that the histogram resembles that abstract and ideal histogram-density curve coined as "The Normal Distribution" that he has heard so much about as freshman at school, and this gave him strong satisfaction as well as some vague assurance that he was on the right path.

But then he got overly enthusiastic, so he attempted to gain even more (illusive) information via the construction of a 50-bin histogram as shown in Figure A.6. This latest histogram turned out to be a complete failure, zigzagging and fluctuating widely, and distorting any potential visual understanding of the data. Some bins in that histogram appear as if they are missing, but this is actually not the case, because these bins are simply empty without any data points falling in, so that their vertical height is 0. This is so because with 50

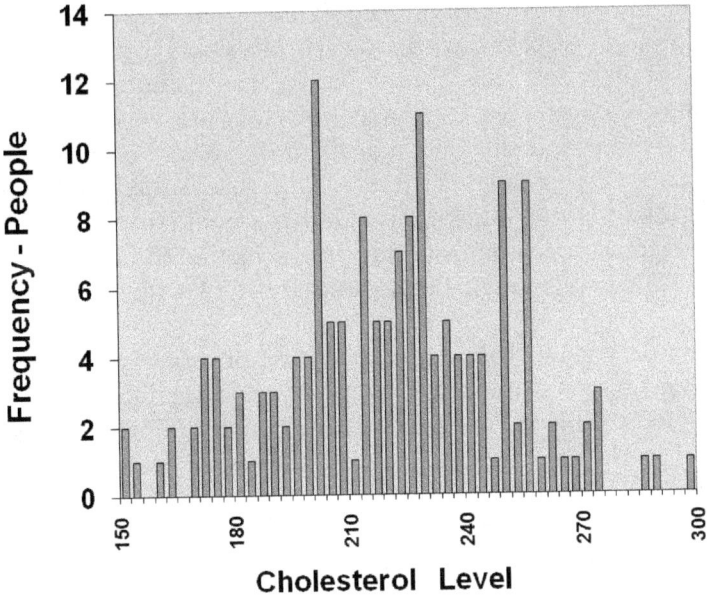

Figure A.6: Histogram of Cholesterol Levels in New York City — 50 Bins

bins, the width of each bin is only three cholesterol units, calculated as $(300 - 150)/50 = (150)/50 = 3$, so that naturally we expect this small size data set of only 160 cholesterol numbers to totally skip some of these narrow and thin bins.

What is the root cause of the problem? Why can't we further scrutinize the data with more refined and even narrower bins? The answer is that for such detailed examination of the cholesterol measurement, it is necessary to have sufficiently many values in the data set, but here we only have 160 values for analysis. Surely, had this New York medical statistical study been based on say tens of thousands of data points, collecting blood samples from that many inhabitants, then each three-cholesterol unit bin would be filled with plenty of data points, even those bins near the 150 and 300 edges, and such 50-bin histogram would look very smooth, round, and decisively clear, while indeed demonstrating a remarkable similarity to the Normal Distribution. With hundreds of thousands of cholesterol samples, we might as well construct a more refined 150-bin histogram, where each bin spans just a singular one-cholesterol unit, and thus obtain an even superior vista of the data. With say a million cholesterol samples, we might as well construct the highly refined 600-bin histogram, where each bin spans a mere 0.25 or 1/4 cholesterol unit, and thus obtain perhaps the ultimate and best vista of the data. Nonetheless, if only very few bins are constructed, so that each bin is wide and thick enough, spanning plenty of horizontal axis range, then even small size data can be reasonably presented with a histogram, as each thick and wide bin would surely have some or at least a few data points falling in, as in the 10-bin histogram of Figure A.5 for this small size cholesterol data with only 160 values in total.

Constructing a histogram for a data set of very big size using truly numerous bins (so that the bins are quite thin having very small width) yields a histogram that appears highly refined and detailed. Thus, when the nature of that big data set itself is decisively of some particular mathematical form (as is the case in so many real-life data), the entire set of the tiny linear roofs on tops of the numerous bins of the histogram appears very smooth and nearly curved, and then an idealized superimposed curve is invented to fit it, and which is what statisticians term "Density Curve". This step transforms the discrete linear histogram into a continuous curve. It should be noted that this step can be properly and formally used only after applying an adjusting multiplicative transformation of the vertical scale, stretching it or squeezing it up or down, to make it more general and independent of the particular size of the data set in use. Figure A.7 is a crude attempt to discover the true or theoretical density curve of New York City cholesterol data by superimposing an imaginary curve along the roofs on tops of the bins. The curve does indeed resemble the Normal Distribution!

As it happened, statisticians have looked into a wide variety of ideal continuous curves built on abstract mathematical formulas, and they have discovered that many of them fit very closely with histograms of real-life data sets having numerous data points and where it is possible to obtain highly refined histograms of very thin bins which appear nearly as curves. Statisticians intentionally design all abstract density curves in such a way so that the total area under the curve from its minimum to its maximum is exactly 1, signifying in a sense 100% of all data points for real-life data histograms. The area of each bin in a histogram is equal to the number of data points falling within it (vertical height), multiplied by the bin's width (horizontal

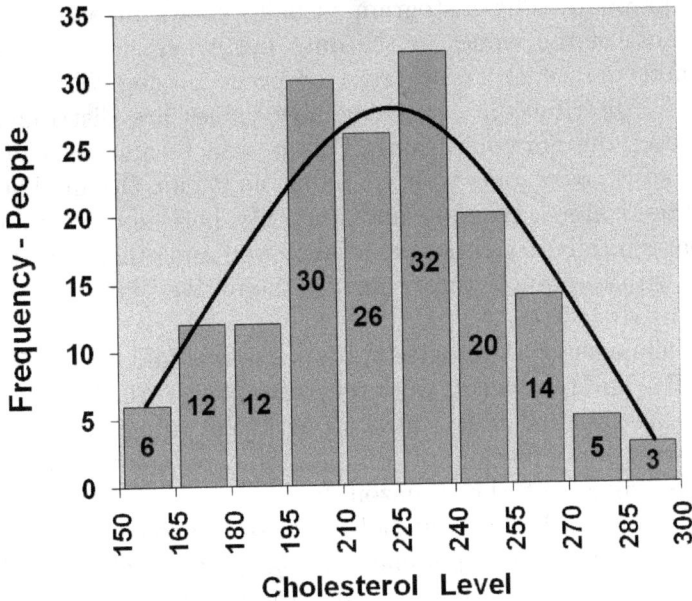

Figure A.7: Crude Approximation of the Density Curve of Cholesterol

length). Therefore, the total area of any histogram is equal to the total number of data points in the data set multiplied by the bin's width. Consequently, in order to transform a refined histogram of a big data set into a continuous probability density function, it is necessary to adjust the scale of the vertical axis by dividing it (i.e. dividing height of each bin) by (I) the total number of data points in the data set, and by (II) the bin's width, and which results in total area of 1 for the adjusted histogram.

For histograms, the horizontal axis stands for the unit of real-life measurement, be it cholesterol levels, weights, heights, kilograms, hours, seconds, days, miles, kilometers, number of accidents, length of objects, dollars, euros, expenses, cost, revenues, tons, centimeters, or inches. The vertical axis (which is the height of any particular bin hovering over a particular range of the horizontal units) stands for the frequency and importance of that particular horizontal range (count of the number of data points falling within the bin of that range).

In essence, what histograms show us is the relative importance of each part of the horizontal axis, from the minimum point to the maximum point. The histogram visually shows in one fell swoop where most of the values in the data fall, where fewer values fall, and where values are quite rare or even missing. This leads to the term "distribution", meaning how values are distributed and spread over the horizontal axis. This is akin to a sudden but brief tropical storm over some wide island in the Pacific Ocean. The storm distributes water unequally and unevenly here and there, leaving some parts totally dry, other parts fairly wet, and other unlucky parts flooded with too much water, hurting the native islanders there and disrupting their lives.

The manager of the novice statistician was highly satisfied with his creative and productive work regarding New York City cholesterol data, so he assigned him a new project concerning the entire data on the tribal population of 160 hunters and gatherers living in a remote village of the Amazon jungle basin in Brazil. He was provided only with the raw cholesterol data shown in Figure A.8. Having already gained experience with his previous cholesterol work, the statistician immediately re-constructed the table of the raw numbers by converting it into a sorted and orderly table as shown in Figure A.9. This clearly demonstrated to him that these tribal hunters are by far healthier than New Yorkers, and which can be confirmed by noticing that approximately the first three columns on the left side are of healthy individuals with ideal levels below the threshold of 200, while the singular column on the right side is with high cholesterol levels of over 200. He then proceeded to draw a proper histogram from the minimum value of 130 to the maximum value of 300, with 17 bins, and where the width of each bin is of 10 cholesterol units, as shown in Figure A.10. This time around he knew not to waste his time experimenting with crude two-bin or three-bin histograms, and also to avoid extremely refined and confusing histograms with too many thin bins.

The statistician was greatly surprised that the tribal histogram was not symmetrical, and he was especially puzzled by the sudden bulge or rise of the bins around the 230–270 range. Reading general literature about the ways of life of Amazonian hunters and their social traditions revealed that each village has a small number of higher-ranked leaders and prominent more assertive male figures. The statistician then speculated on the hypothesis that after each successful hunt these higher ranked individuals aggressively demand

268.9	166.5	164.5	145.9
174.8	197.7	167.3	161.1
262.5	168.0	181.3	298.9
173.0	169.3	167.5	169.3
146.7	167.3	177.5	164.3
162.3	209.0	157.6	154.1
169.8	157.1	161.0	230.1
161.6	221.9	263.5	199.2
167.8	198.4	174.3	154.4
153.9	164.1	243.3	148.2
180.6	183.4	157.6	255.7
201.3	132.0	173.8	253.3
179.4	256.3	204.3	247.8
232.6	151.0	142.9	186.2
250.3	200.1	191.2	292.4
212.6	157.0	169.7	248.5
150.8	161.3	151.7	251.2
169.9	178.4	145.9	246.6
243.9	130.5	248.2	173.0
155.4	167.2	195.5	167.1
163.3	246.9	253.3	229.4
153.2	258.3	164.2	199.4
165.2	237.2	178.3	163.6
168.8	141.0	247.3	187.5
139.0	254.5	193.6	180.2
188.4	144.7	267.6	156.4
170.8	283.9	239.8	257.4
165.9	180.6	233.8	190.2
141.1	165.8	266.7	194.2
244.7	163.9	256.7	227.4
183.2	169.1	233.9	156.5
218.3	179.5	149.5	212.8
159.9	155.6	171.5	157.5
181.5	252.3	145.5	234.4
249.5	207.3	178.5	182.9
253.3	170.2	167.1	228.8
253.5	171.4	276.2	154.4
158.3	153.0	179.3	239.3
166.3	288.3	243.8	157.1
144.8	154.1	151.3	169.3

Figure A.8: Cholesterol Levels in Brazil Amazon Tribe — Raw Data

and actually obtain the fatter, oily, and more delicious parts of the animal caught, and such continuously bad diet adversely affects their cholesterol level. Hence the histogram in Figure A.10 could be interpreted as the fusion of two distinct Normal-like histograms, one for most of the lower-ranked villagers centered around the healthy

130.5	161.6	178.3	233.8
132.0	162.3	178.4	233.9
139.0	163.3	178.5	234.4
141.0	163.6	179.3	237.2
141.1	163.9	179.4	239.3
142.9	164.1	179.5	239.8
144.7	164.2	180.2	243.3
144.8	164.3	180.6	243.8
145.5	164.5	180.6	243.9
145.9	165.2	181.3	244.7
145.9	165.8	181.5	246.6
146.7	165.9	182.9	246.9
148.2	166.3	183.2	247.3
149.5	166.5	183.4	247.8
150.8	167.1	186.2	248.2
151.0	167.1	187.5	248.5
151.3	167.2	188.4	249.5
151.7	167.3	190.2	250.3
153.0	167.3	191.2	251.2
153.2	167.5	193.6	252.3
153.9	167.8	194.2	253.3
154.1	168.0	195.5	253.3
154.1	168.8	197.7	253.3
154.4	169.1	198.4	253.5
154.4	169.3	199.2	254.5
155.4	169.3	199.4	255.7
155.6	169.3	200.1	256.3
156.4	169.7	201.3	256.7
156.5	169.8	204.3	257.4
157.0	169.9	207.3	258.3
157.1	170.2	209.0	262.5
157.1	170.8	212.6	263.5
157.5	171.4	212.8	266.7
157.6	171.5	218.3	267.6
157.6	173.0	221.9	268.9
158.3	173.0	227.4	276.2
159.9	173.8	228.8	283.9
161.0	174.3	229.4	288.3
161.1	174.8	230.1	292.4
161.3	177.5	232.6	298.9

Figure A.9: Cholesterol Levels in Brazil Amazon Tribe — Sorted Data

170 level, and another for the fewer higher-ranked leaders centered around the unhealthy 250 level.

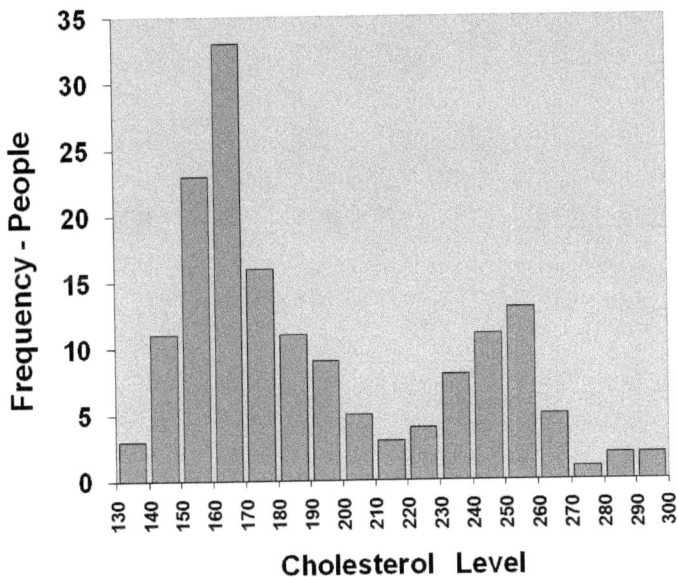

Figure A.10:   Histogram of the Cholesterol Levels in Brazil Amazon Tribe

Appendix B

Conceptual Explanation
of Powers and Logarithms

When we add numbers, resultant values increase steadily, but when we multiply numbers by each other, it leads to much greater and more rapid increase in values. This is so because multiplication signifies repeated additions, over and over again. For example, $6 + 5$ is just 11, but 6×5 is actually $5 + 5 + 5 + 5 + 5 + 5$ and which is 30. Naturally, the next higher level of rapid growth in values is repeated multiplications, such as for example $5 \times 5 \times 5$ which is 125, or as in $10 \times 10 \times 10 \times 10 \times 10 \times 10$ which is a million.

It was much more convenient and concise for us to write the long sum of six fives $5 + 5 + 5 + 5 + 5 + 5$ as simply 6×5, or to write the longer sum of nine twos $2 + 2 + 2 + 2 + 2 + 2 + 2 + 2 + 2$ as simply 9×2. In the same vein, repeated multiplications are written more conveniently and concisely by indicating only two values; one value is that of the particular number about to be multiplied repeatedly, and which is called the "**base**", and another value indicating by how many times the base should be multiplied by itself, and which is called the "**exponent**". Suppose a base number B is being multiplied by itself N number of times, then the resultant quantity of such multiplicative process is called and expressed as "the Nth power of the base B." To be precise, the term "exponent" refers only to the number of times a given base B is being multiplied by itself, while the term "**power**" refers to the final value of the process after the base had been multiplied by itself several times. This process leads to a powerful and extremely rapid increase in values, hence the term "power".

The convention for the symbols and notations here is such that the base B is indicated by writing it as a regular size, while the exponent N is indicated by writing it with a much smaller size and placing it on the top-right corner of the base B, namely as superscript, so that powers are denoted as **BASE**$^\textbf{EXPONENT}$ or even more concisely as B^N. This is coined as in the phrase "B raised to the Nth power"; or as "B to the power N"; and even very briefly as "B to the N". When N is 2, this is called a "square", or "the square of B", as if referring to two-dimensional area on the plane. When N is 3, this is called a "cube", or "the cube of B", as if referring to three-dimensional volume in space.

Let us give some concrete examples:

3^4 which is $3 \times 3 \times 3 \times 3$ or simply 81

5^2 which is 5×5 or simply 25

5^3 which is $5 \times 5 \times 5$ or simply 125

2^4 which is $2 \times 2 \times 2 \times 2$ or simply 16

10^3 which is $10 \times 10 \times 10$ or simply 1000

1^7 which is $1 \times 1 \times 1 \times 1 \times 1 \times 1 \times 1$ or simply 1

0^3 which is $0 \times 0 \times 0$ or simply 0

Surely when the exponent is 1, there is actually no multiplication involved whatsoever, because the conceptual interpretation of it is that of the value of the base, considered just once, as if standing by itself, therefore, $3^1 = 3$, $5^1 = 5$, $100^1 = 100$, and $77^1 = 77$.

When the exponent is 0 the result is 1, and this is so regardless of the value of the base (unless perhaps when the base is also 0), therefore, $3^0 = 1$, $5^0 = 1$, $100^0 = 1$, and $77^0 = 1$. Most new students of mathematics are at a complete loss understanding this fact as it seems to have no explanation, but this feature of powers is essential in preventing contradictions, and it expresses something very profound in this field of powers, as it harmonizes with several formal rules of powers. This feature is also essential is guaranteeing that the graph of the Exponential Function $f(x) = \text{BASE}^x$ is smooth, continuous, and without any gaps or tears. This graph shall be discussed here in a later paragraph.

When the exponent is a fraction in the form of (one)/(integer) or $1/R$, R being some particular integer, so that the structure of the power is $\text{BASE}^{(1/R)}$, the result is defined as the Rth root of the base. If R is 2 then it's the square root of the base. If R is 3 then it's the cube root of the base. If $R > 3$, then it's the Rth root of the base. For example, $9^{(1/2)}$ = the square root of $9 = 3$. The square root of 9 is simply the answer to the riddle asking to find a number, called S, such that when that elusive number is multiplied by itself two times the result is the number 9, and this is denoted by writing the riddle as $S \times S = 9$, and therefore surely, that elusive number is simply 3 since 3×3 is 9. As another example, $8^{(1/3)}$ = the cube root of $8 = 2$. The cube root of 8 is simply the answer to the riddle asking to find a number, called C, such that when that elusive number is multiplied by itself three times the result is the number 8, and this is denoted by writing the riddle as $C \times C \times C = 8$, and therefore surely, that elusive number is simply 2 since $2 \times 2 \times 2$ is 8. In the same vein and in extreme generality, B to the power $1/N$, namely $B^{(1/N)}$, is simply the Nth root of B, which is that elusive number, called T, such that when multiplied by itself N times the result is the number B, and this is denoted by writing the riddle as $[T \times T \times T \times \ldots N \text{ times} \ldots \times T \times T \times T] = B$.

When the exponent is a negative number, so that the structure of the power is $\text{BASE}^{(-N)}$, the result is defined as the number 1 divided by the positive power, namely $1/\text{BASE}^{(+N)}$. For example, $3^{(-2)} = 1/3^{(+2)} = 1/(3 \times 3) = 1/9 = 0.11111$. As another example, $10^{(-3)} = 1/10^{(+3)} = 1/(10 \times 10 \times 10) = 1/(1000) = 0.001$.

Figure B.1 depicts the Exponential Function $f(\text{Exponent}) = 2^{\text{EXPONENT}}$ for a variety of exponent values ranging from -4 to $+4$. The horizontal axis represents the Exponent, while the vertical axis represents the value of 2^{EXPONENT} for each exponent value. For example, when the exponent is 3 on the horizontal axis, the power value of 2^3 being 8 is symbolized by the height of the vertical axis. When the exponent is 1 on the horizontal axis, the power value of 2^1 being 2 is symbolized by the height of the vertical axis. When the exponent is 0 on the horizontal axis, the power value of 2^0 being 1 is symbolized by the height of the vertical axis. When the exponent is -1 (negative one) on the horizontal axis, the power value of 2^{-1} being 1/2 or 0.5 is symbolized by the height of the vertical axis. When the exponent is -2 (negative two) on the horizontal axis, the

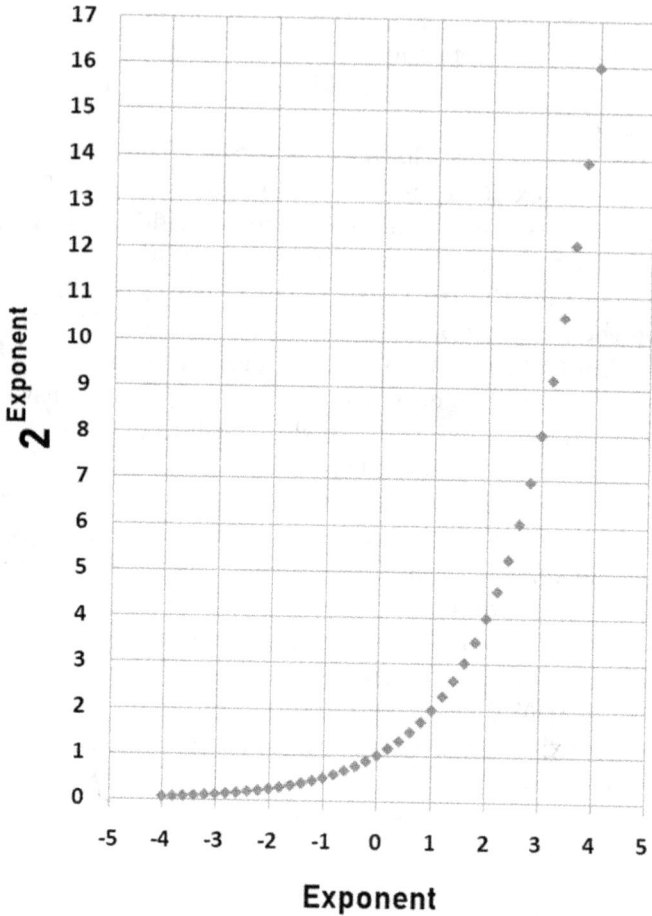

Figure B.1:   The Exponential Function of 2 to the Power of Various Exponents

power value of 2^{-2} being $1/(2 \times 2)$ namely $1/4$ or 0.25, is symbolized by the height of the vertical axis.

One well-known rule of powers is that for an exponent composed of two multiplicands such as $R \times Q$, the value of the entire power can be calculated in stages, first by evaluating the base to exponent R, and then that result is considered as a new base, and which leads to the next evaluation by raising it to exponent Q. The rule is concisely denoted by the equation $B^{(R \times Q)} = [B^{(R)}]^{(Q)}$. For example, $B^{(2 \times 4)} = B^{(8)} = B \times B \times B \times B \times B \times B \times B \times B = [B \times B] \times [B \times B] \times [B \times B] \times [B \times B] = [B \times B]^{(4)} = [B^{(2)}]^{(4)}$. Applying this rule to any fractional exponent of the form $B^{(N/D)}$, with integer N denoting the numerator, and with integer D denoting the denominator, leads to another rule via basic algebraic manipulations as in:

$$B^{(N/D)} = B^{((N/1)\times(1/D))} = B^{((1/D)\times(N/1))} = [B^{(1/D)}]^{(N/1)} = [B^{(1/D)}]^{(N)}$$

Hence, the straightforward interpretation of $B^{(N/D)}$ where N and D are integers, is simply a two-stage evaluations process, where we first calculate the Dth-root of B, leading to a new base, and then secondly we take that new base to the power of exponent N. As an example, $8^{(5/3)} = [8^{((1/3)\times(5/1))}] = [8^{(1/3)}]^{(5)} = [\text{3rd root of 8}]^{(5)} = [2]^{(5)} = 32$, namely first we evaluate the cube root of 8 which is 2, followed secondly by the evaluation of 2 to the exponent 5, and which is 32. In another example, $2^{(1.6)} = 2^{(8/5)} = [2^{((1/5)\times(8/1))}] = [2^{(1/5)}]^{(8)} = [\text{5th root of 2}]^{(8)} = [1.14869835]^{(8)} = 3.03143313$; and this could be seen in the above graph as the point approximately 1.6 horizontal units to the right and 3 vertical units upward.

Another well-known and highly intuitive rule of powers is that the multiplication of two distinct powers sharing the same base B, namely, a power of base B to the exponent M, multiplied by the power of base B to the exponent N, can be evaluated and expressed neatly in one fell swoop, fusing it into a singular power of the base B to the sum the exponents, namely $M + N$. The rule is concisely denoted by the equation $[B^M] \times [B^N] = B^{(M+N)}$. For example, for the product of two distinct powers $[7^3]$ and $[7^5]$ with a common base 7, we easily obtain the concise result of $[7^3] \times [7^5] = [7 \times 7 \times 7] \times [7 \times 7 \times 7 \times 7 \times 7] = [7 \times 7 \times 7 \times 7 \times 7 \times 7 \times 7 \times 7] = 7^8 = 7^{(3+5)}$.

Another well-known and highly intuitive rule of powers is that the ratio of two powers, both sharing a common base B, such as a power of base B to the exponent N, divided by the power of base B to the exponent M, can be evaluated in one fell swoop and fused into a singular power of the base B to the difference $N - M$, namely with the exponent value of the difference between N and M. The rule is concisely denoted by the equation $[B^N]/[B^M] = B^{(N-M)}$. For example $[2^6]/[2^4] = [2 \times 2 \times 2 \times 2 \times 2 \times 2]/[2 \times 2 \times 2 \times 2] = [2 \times 2] = 2^2 = 2^{(6-4)}$.

Exponential notation and the general power form of numbers help us in expressing and representing extremely large as well as extremely small numbers in a very convenient and concise manner. For example,

1000000000000 can be represented as 1×10^{12}, whereas 0.000000003 can be represented as 3×10^{-9}. All this facilitates readability, provides for easy writing, saves time, and brings elegance to numbers and mathematics.

While lecturing on the topic of this book at various universities during 2023 and 2024, this author frequently posed to a group of mathematicians the question regarding the true power value of base zero to the exponent of another zero, namely 0^0. Surprisingly, there was no consensus among these highly esteemed professors of mathematics with years or even decades of academic research and teaching experience, and immediately the issue divided them into two opposing camps, with one camp claiming that the value is one, namely $0^0 = 1$, and another camp claiming that the value is zero, namely $0^0 = 0$. Several mathematicians on the other hand, addressed the issue as the result of two simultaneous limit processes denoted as in: limit $X \to 0$ and limit $Y \to 0$ of X^Y, and which depends on the specifics and details of these limiting processes themselves, namely the fast or slow rate of convergence to zero experienced by X compared to that experienced by Y.

Karl Schwarzschild (1873–1916) was a German physicist. While fighting in World War I at the age of about 40, he attempted to draw further conclusions, find new consequences, and provide deeper analysis of Einstein's theory of general relativity. In 1916, while still serving in the German army, he then wrote about the possibility of the existence of cosmological black holes as a consequence of relativity, and published this in an article titled in German "Über das Gravitationsfeld eines Massenpunktes nach der Einsteinschen Theorie" which translates into English as "On the Gravitational Field of a Mass Point According to Einstein's Theory". Schwarzschild, who died soon after his publication in 1916 on the Russian Front from either an infection or from his own illness unrelated to the war, provided the first exact solution to the Einstein field equations of general relativity, for the limited case of a single spherical non-rotating mass, leading to the idea of black holes. In an article in 2022 published on Hal open science platform, it was claimed that perhaps Schwarzschild's derivation on black hole is based on an irregular change of variable associated with the hypothesis that zero raised to the zero power is equal to one, namely that supposedly the entire original study on black holes is based on the assumption that $0^0 = 1$. Readers might find the whole discussion fascinating, highlighting the importance and impact of mathematical concepts and its purely abstract knowledge on our understanding of the physical world around us. Is the Creator or Mother Nature truly constrained and greatly limited by mathematical facts, and without the ability to create wild, unconstraint, and totally innovative things in the physical world without first asking for a formal approval from the Goddess of Mathematics, always waiting for her positive and enthusiastic reply before taking any further action? The answer here is decisively in the affirmative according to two essential Renaissance scientists who directly paved the way for Isaac Newton's *Philosophiae Naturalis Principia Mathematica*. One quote attributed to Galileo Galilei states that "Mathematics is the language in which God wrote the universe." Another quote attributed to Johannes Kepler states that "Geometry was co-eternal with the Divine Mind prior to the birth of all things." Science forever bows with deep respect and humility to mathematics, always seeking confirmation, permission, validation, and approval before daring to discover any new laws or pattern, and even before starting to define its own constructs, entities, dimensions, scales, units, and the setup of inquiry. The alternative heretical philosophical approach is that mathematics is not a prerequisite for science, but rather an aid and facilitator, as in the analogy with shoes which help us to walk. We can walk without any shoes outside our house, slowly and with risks, but when walking with shoes, we can wonder around confidently, safely, and go faster and farther.

The concept of power and exponents in mathematics, can be traced back to ancient civilizations, notably with the work of the mathematical genius Archimedes of ancient Greece, then also with the late Middle Ages mathematician Nicolas Chuquet in France around 1484, and finally with Michael Stifel in Germany during the Renaissance Era who published a book titled *Arithmetica Integra* in 1544, which contained innovative mathematical notations for powers and exponents, including two power rules, and coining the term "exponent". This was followed by the Scottish mathematician James Hume in UK around 1639 who is given credit for introducing the modern exponential notations, and more notably so with the French philosopher René Descartes with his publication of *La Géométrie* in 1637. Descartes was one of the key figures in developing the notation and understanding of exponents. He introduced modern exponential notation for powers and broke with the Greek tradition of associating powers with geometric concepts, such as with an area for squares, and with a volume for cubes, and so on, treating square, cubes, and higher powers as possible lengths of line segments. The systematic use of exponents and their properties became formalized and then further polished by later mathematicians, including John Napier of Scotland, who introduced the concept of the logarithms, and Isaac Newton of England, who explored powers for his work on calculus in his quest to develop Classical Mechanics as the basis of the deepest, the most exact, and unifying explanation of all physical phenomenon.

The concept of the **logarithm** cannot truly be thought of as the opposite notion or reverse operation of the idea of powers, rather only roots, square roots, and cube roots can be thought of as the polar opposites of powers. For powers, namely the format of **BASE^{EXPONENT} = Result**, that result is decisive as it always points to one unique and known value, but logarithms, on the other hand, should be thought of as a riddle or a puzzle, asking for some illusive or authentic number that may serve as the exponent, and whose possible (but not assured) answer is a value obtained by knowing the base value as well as the resultant quantity of some particular power arrangement, while the exponent itself (the logarithm) is not known. The logarithm of the power arrangement **BASE$^{???}$ = Result**, is denoted by "???". Indeed, the logarithm can be thought of as that unknown exponent to which the known base is raised to yield that known resultant power. Logarithm is denoted by $\text{LOG}_{\textbf{BASE}}(\text{Result})$, or more concisely as $\text{LOG}_{\textbf{B}}(N)$. It's simply a riddle, asking "what exponent do we need for the base B in order for such power arrangement to result in N?" For example, $\text{LOG}_2(8)$ is simply a question, asking what exponent do we need for the base 2, so that 2 to the power of that illusive exponent yields 8. The answer is clearly the number 3, because two multiplied by itself three times yields 8. In other words, since $2^3 = 2 \times 2 \times 2 = 8$, therefore $\text{LOG}_2(8) = 3$. Clearly, the concept of the logarithm is based on the concept of powers and nothing else. In a sense, the only difference between powers and logarithms is which of the 3 values is being left unknown and has to be calculated or found. For powers, the equation $2^3 = X$ informs us about the base, which is known to be 2, it also informs us about exponent, which is known to be 3, but it leaves out the result (the power) as unknown, and which could be easily calculated as $2 \times 2 \times 2$ yielding the singular value 8. For logarithms, the equation $LOG_2(8) = X$ informs us about the base, which is known to be 2, it also informs us about the result (the power), which is known to be 8, but it leaves out the exponent as unknown, and which could only be imagined or guessed. Clearly, the exponent needed here to go from the base of 2 and to arrive at the power value of 8 is evidently 3.

Let us look at other examples:

$LOG_3(81) = 4$

$LOG_5(25) = 2$

$LOG_5(125) = 3$

$LOG_2(16) = 4$

$LOG_{10}(1000) = 3$

$LOG_{10}(100) = 2$

$LOG_{10}(10) = 1$

$LOG_{10}(1) = 0$

$LOG_{77}(77) = 1$

$LOG_{77}(1) = 0$

$LOG_{25}(5) = 0.5$

The fact that base 1 to the power of any exponent whatsoever always yields 1, as in the examples $1^{35} = 1$, $1^7 = 1$, $1^3 = 1$, $1^1 = 1$, $1^0 = 1$, $1^{-1} = 1$, $1^{-5} = 1$, and $1^{(1/4)} = 1$, leads to uncertainty and confusion when asking about the value of $LOG_1(1)$. Therefore, asking about the value of $LOG_1(1)$ is a puzzle that leads to uncertainty and confusion, as it could be any number. The same confusing fate befalls base 0 raised to any power (except 0 itself perhaps), and which always or nearly always yields 0, as in the examples $0^{47} = 0$, $0^6 = 0$, $0^1 = 0$, and $0^{(1/3)} = 0$, while negative exponents for base zero such as 0^{-1} and 0^{-5} are undefined or rather infinite since we are not allowed to divide by zero. Therefore, asking about the value $LOG_0(0)$ is a riddle that leads to uncertainty and confusion, as it could be any non-zero positive number. Also, asking about the value of $LOG_0(7)$ is a puzzle without any possible answer, since no exponent whatsoever for base zero could be found which yields 7. In addition, since powers of any positive base $B > 0$ always yield non-negative results, and regardless of the value of the exponent, asking the value of the logarithm of a negative number is not allowed, and therefore all expressions such as $LOG_{10}(-100)$, $LOG_2(-8)$, $LOG_5(-25)$, and so forth, where the quantity in the parenthesis is negative, are all forbidden puzzles that we are not allowed to ask. The only way negative values are allowed in logarithms is the rare and odd case when we assign the base itself to be negative, for example, when the base itself is -2 and which allows us to ask for the value of $LOG_{(-2)}(-8)$, pointing to the answer $+3$. Another rare and odd example is the case of the base being -5, which allows us to ask for the value of $LOG_{(-5)}(-125)$, pointing to the answer $+3$.

Two logarithmic-power rules signify their authentic inverse relationship:

$$LOG_{BASE} (BASE^{NUMBER}) = NUMBER$$

$$BASE^{LOG_{BASE}NUMBER} = NUMBER$$

Logarithmic functions are the inverses of exponential functions, and undo the exponential functions. Exponential functions are the inverses of logarithmic functions, and undo the logarithmic functions.

The following equations are three very practical rules of logarithms:

$$LOG_B(N \times M) = LOG_B(N) + LOG_B(M)$$

$$LOG_B(N/M) = LOG_B(N) - LOG_B(M)$$

$$LOG_B(X^{\mathbf{R}}) = R \times LOG_B(X)$$

Glossary of Frequently Used Abbreviations

CLT   Central Limit Theorem

MCLT Multiplicative Central Limit Theorem

OOM   Order of Magnitude — LOG(Maximum/Minimum)

POM Physical Order of Magnitude — Maximum/Minimum

CPOM Core Physical Order of Magnitude — $P_{90\%}/P_{10\%}$

GLORQ The General Law of Relative Quantities

SSD   Sum of Squared Deviation Measure

IPOT Integral Powers of Ten

Bibliography

Andrews G. (1976). *The Theory of Partitions*. Cambridge University Press.

Benford F. (1938). "The Law of Anomalous Numbers". *Proceedings of the American Philosophical Society*, 78, 551.

Buck B., Merchant A., Perez S. (1992). "An Illustration of Benford's First Digit Law Using Alpha Decay Half Lives". *European Journal of Physics*, 14, 59–63.

Carslaw C. (1988). "Anomalies in Income Numbers: Evidence of Goal Oriented Behavior". *The Accounting Review*, 1988, 321–327.

Deckert J., Myagkov M., Ordeshook P. (2011). "Benford's Law and the Detection of Election Fraud". *Political Analysis* 19(3), 245–268.

Durtschi C., Hillison W., Pacini C. (2004). "The Effective Use of Benford's Law to Assist in Detecting Fraud in Accounting Data". *Auditing: A Journal of Forensic Accounting*, 1524-5586/Vol. V, 17–34.

Gaines J. B., Cho K. W. (2007). "Breaking the (Benford) Law: Statistical Fraud Detection in Campaign Finance". *The American Statistician*, 61(3), 218–223.

Hamming R. (1970). "On the Distribution of Numbers". *Bell System Technical Journal*, 49(8), 1609–1625.

Kossovsky A. E. (2012). "Towards A Better Understanding of the Leading Digits Phenomena". City University of New York. http://arxiv.org/abs/math/0612627

Kossovsky A. E. (December 2012). "Statistician's New Role as a Detective — Testing Data for Fraud". http://revistas.ucr.ac.cr/index.php/economicas/article/view/8015

Kossovsky A. E. (2013). "On the Relative Quantities Occurring within Physical Data Sets". http://arxiv.org/ftp/arxiv/papers/1305/1305.1893.pdf

Kossovsky A. E. (August 2014). *Benford's Law: Theory, the General Law of Relative Quantities, and Forensic Fraud Detection Applications.* World Scientific Publishing Company. ISBN: 978-981-4583-68-8.

Kossovsky A. E. (2015). "Random Consolidations and Fragmentations Cycles Lead to Benford's Law". https://arxiv.org/abs/1505.05235

Kossovsky A. E. (March 2016). "Prime Numbers, Dirichlet Density, and Benford's Law". https://arxiv.org/abs/1603.08501

Kossovsky A. E. (May 2016). "Arithmetical Tugs of War and Benford's Law". http://arxiv.org/abs/1410.2174

Kossovsky A. E. (June 2016). "Exponential Growth Series and Benford's Law". http://arxiv.org/abs/1606.04425

Kossovsky A. E. (February 2019). "Quantitative Partition Models and Benford's Law". https://arxiv.org/abs/1606.02145

Kossovsky A. E. (April 2019). *Studies In Benford's Law: Arithmetical Tugs of War, Quantitative Partition Models, Prime Numbers, Exponential Growth Series, and Data Forensics.* Kindle Direct Publishing, Amazon. https://www.amazon.com/dp/172928325X

Kossovsky A. E. (August 2020). *The Birth of Science, Kepler's Celestial Data Analysis, Galileo's Terrestrial Experiments, and Newton's Grand Synthesis.* Springer Nature Publishing. https://www.amazon.com/Birth-Science-Springer-Praxis-Books/dp/3030517438/, https://link.springer.com/book/10.1007/978-3-030-51744-1

Kossovsky A. E. (May 2021). "On the Mistaken Use of the Chi-Square Test in Benford's Law". https://www.mdpi.com/2571-905X/4/2/27

Kossovsky A. E., Lawton W. (August 2023). "A Mathematical Analysis of Benford's Law and its Generalization". https://arxiv.org/pdf/2308.07773.pdf

Kossovsky A. E., Miller S. (December 2020). "Report on Benford's Law Analysis of 2020 Presidential Election Data". https://web.williams.edu/Mathematics/sjmiller/public_html/KossoskyMiller_FinalBenford Analysis.pdf

Leemis L., Schmeiser B. W., Evans D. L. (November 2000). "Survival Distributions Satisfying Benford's Law". *The American Statistician,* 54(4), 236–241.

Leuenberger C., Engel H.-A. (2003). "Benford's Law for the Exponential Random Variables". *Statistics and Probability Letters,* 63(4), 361–365.

Miller S. (June 2008). "Chains of Distributions, Hierarchical Bayesian Models and Benford's Law". http://arxiv.org/abs/0805.4226

Miller S. *et al.* (December 30, 2013). "Benford's Law and Continuous Dependent Random Variables." http://arxiv.org/pdf/1309.5603.pdf

Miller S., Joseph I., Frederick S. (2015). "Equipartitions and a Distribution for Numbers: A Statistical Model for Benford's Law". Williams College.

Newcomb S. (1881). "Note on the Frequency of Use of the Different Digits in Natural Numbers". *American Journal of Mathematics*, 4, 39–40.

Pinkham R. (1961). "On the Distribution of First Significant Digits". *The Annals of Mathematical Statistics*, 32(4), 1223–1230.

Raimi A. R. (1969). "The Peculiar Distribution of First Digit". *Scientific America*, 109–115.

Raimi A. R. (1976). "The First Digit Problem". *American Mathematical Monthly*.

Raimi A. R. (1985). "The First Digit Phenomena Again". *Proceedings of the American Philosophical Society*, 129(2), 211–219.

Ross A. K. (2011). "Benford's Law, A Growth Industry". *The American Mathematical Monthly*, 118(7), 571–583.

Sambridge M., Tkalcic H., Arroucau P. (October 2011). "Benford's Law of First Digits: From Mathematical Curiosity to Change Detector". *Asia Pacific Mathematics Newsletter*.

Sambridge M., Tkalcic H., Jackson A. (2010). "Benford's Law in the Natural Sciences". *Geophysical Research Letters*, 37(22), L22301.

Saville A. (2006). "Using Benford's Law to detect data error and fraud: An examination of companies listed on the Johannesburg Stock Exchange". *South African Journal of Economics and Management Sciences*, 9(3), 341–354. http://repository.up.ac.za/handle/2263/3283

Shao L., Ma B.-Q. (May 6, 2010a). "The Significant Digit Law in Statistical Physics". http://arxiv.org/abs/1005.0660

Shao L., Ma B.-Q. (May 10, 2010b). "Empirical Mantissa Distributions of Pulsars". *Astroparticle Physics*, 33, 255–262. http://arxiv.org/abs/1005.1702

Varian H. (1972). "Benford's Law". *The American Statistician*, 26(3).

Patent: The U.S. Patent Office # 9,058,285.
Inventor: Alex Ely Kossovsky.

Date Granted: June 16, 2015.
http://www.google.com/patents/US20140006468

Titled: "Method and system for Forensic Data Analysis in fraud detection employing a digital pattern more prevalent than Benford s Law".

Index